FATIGUE CRACK GROWTH

30 Years of Progress

FATIGUE CRACK GROWTH
30 Years of Progress

Proceedings of a Conference on Fatigue Crack Growth
Cambridge, UK, 20 September 1984

Edited by

R. A. SMITH

University of Cambridge, Cambridge, UK

PERGAMON PRESS

OXFORD · NEW YORK · TORONTO · SYDNEY · FRANKFURT

U.K.	Pergamon Press Ltd., Headington Hill Hall, Oxford OX3 0BW, England
U.S.A.	Pergamon Press Inc., Maxwell House, Fairview Park, Elmsford, New York 10523, U.S.A.
CANADA	Pergamon Press Canada Ltd., Suite 104, 150 Consumers Road, Willowdale, Ontario M2J 1P9, Canada
AUSTRALIA	Pergamon Press (Aust.) Pty. Ltd., P.O. Box 544, Potts Point, N.S.W. 2011, Australia
FEDERAL REPUBLIC OF GERMANY	Pergamon Press GmbH, Hammerweg 6, D-6242 Kronberg, Federal Republic of Germany
JAPAN	Pergamon Press Ltd., 8th Floor, Matsuoka Central Building, 1-7-1 Nishishinjuku, Shinjuku-ku, Tokyo 160, Japan
BRAZIL	Pergamon Editora Ltda., Rua Eça de Queiros, 346, CEP 04011, São Paulo, Brazil
PEOPLE'S REPUBLIC OF CHINA	Pergamon Press, Qianmen Hotel, Beijing, People's Republic of China

First edition 1986

Library of Congress Cataloging in Publication Data
Conference on Fatigue Crack Growth (1984: Cambridge, Cambridgeshire)
Fatigue crack growth.
1. Materials — Fatigue — Congresses. I. Smith, R. A.
II. Title.
TA418.38.C66   1984   620.1'123   85-28362

British Library Cataloguing in Publication Data
Fatigue crack growth: 30 years of progress: proceedings of a conference on fatigue crack growth, Cambridge, UK, 20 September 1984.
1. Materials—Fatigue
I. Smith, R. A. (Roderick Arthur)
620.1'23   TA418.38
ISBN 0-08-032547-5

Printed in Great Britain by A. Wheaton & Co. Ltd., Exeter

Contents

Introduction

These papers are the proceedings of a Symposium held in September 1984 which served two purposes. Three of the outstanding figures of the fatigue world - Peter Forsyth, Norman Frost and Gerry Smith - have recently retired. Their careers spanned a period during which our understanding of the fatigue crack growth process has increased enormously. It was therefore felt that a review of this period would be of great interest and would serve to mark their contributions.

A short summary of the careers of the three principals is given, together with their own choice of their five most interesting and/or significant works. What is hard to express in print, and is perhaps inappropriate in a formal volume, is the warmth and affection with which these men are held in the fatigue, and wider, communities. This volume aims to express our appreciation of their activities.

Two of the papers by Fleck and Young, were not presented at the Symposium, but have been added to provide balance to the volume. The first two papers are historical reviews, by Smith on the development in Fatigue Crack Growth and by Marsh and Smith on Fatigue Testing Equipment - improvements of which have made possible many advances. Although there is, quite naturally, an emphasis on the British scene, a glance at the contents and references of these and the subsequent papers, should serve to dispel any charges of insularity from a wider international view. Knott deals with the modelling of the crack growth process, whilst Lindley and Nix review metallurgy - vital aspects of fatigue crack growth. Although the work of the last thirty years has seen a great simplification of the techniques of approaching fatigue crack propagation (largely through the use of the alternating stress intensity factor), research is still being conducted at an ever increasing rate. In many cases, this work is in the nature of dotting the i's, but Fleck's paper on the complications of fatigue crack growth seeks to identify which problems are significant and which are second order. The paper by Beevers and Carlson on the important topic of thresholds, might be viewed as representative of the fine detail of current research. Although the vast majority of fatigue problems occur in metals, Young's paper on non-metallics serves to remind us of the increasing use of such materials and the increasing sophistication of their design. Harrison's paper on Damage Tolerant Design reviews the approaches now possible to the design of defective (cracked) structures and how the advances in fracture mechanics over our period of review have been fed into the industrial design scene. The final paper by Tomkins attempts to draw the threads together and speculates on the future of cracks and defects.

The Comet accidents of thirty years ago involved the mysterious catastropic disintegration of aircraft in stable flight. By a curious coincidence of timing, as these papers were being prepared an Air India Jumbo disappeared into the Atlantic off the Irish coast. Strong, albeit circumstantial, evidence points to terrorist activity, but no firm proof of this theory has emerged. This accident was rapidly followed by the worst ever loss of life on a single aircraft - a Japanese Jumbo, and a serious fire on take off to a British 737. Both the latter accidents were caused by some form of mechanical and structural failure involving fatigue. A sad reminder of the technological difficulties of applying our fatigue knowledge to real components.

I have enjoyed stimulating discussions with Norman Frost, Peter Forsyth and Gerry Smith during the preparations for our Symposium and would like to record my thanks to them. Production of camera-ready copy (with its attendent defects!) does have the advantage of requiring close collaboration with the authors, which I have thoroughly enjoyed. Finally, to many people in Cambridge University Engineering Department, particularly my research students Dr Wilf Nixon, James Cooper, Chou Sheng Shin and Grant Leaity, the photographer John Read, Sheila Owen of the Drawing Office, and last, but by no means least, Rosalie Orriss for her help in producing the manuscripts, I record my grateful thanks.

R A Smith

Cambridge University
Engineering Department

Biographical Notes

N. E. FROST, CBE

Norman Frost served an engineering apprenticeship with Rolls-Royce, Derby, obtaining a London Mechanical Engineering degree in the process. After four years in Rolls-Royce research laboratories, in 1948 he joined the National Physical Laboratory working on high temperature materials problems. In 1952 he transferred to the Mechanical Engineering Research Laboratory (latterly the National Engineering Laboratory, East Kilbride), of which, after several promotions over the years, he was appointed Deputy Director in 1977.

His technical interest has always been centred around engineering aspects of fatigue, particularly for steels, and he has published some 60 papers on notch effects, crack growth rates and threshold levels for fatigue crack growth.

"A relation between the critical alternating stress for propagation and crack length for mild steel", Proc. Instn. Mech. Engrs. 1959 <u>173</u> 811

"Non-propagating cracks in Vee-notched specimens subject to fatigue loading", Aero. Quart. 1957 <u>8</u> 1

"Significance of non-propagating cracks in the interpretation of notched fatigue data", J. Mech. Eng. Sci. 1961 <u>3</u> 299

"A fracture mechanics analysis of fatigue crack growth data for various materials", with L P Pook and K Denton, Eng. Fracture Mech. 1971 <u>3</u> 109

"Metal fatigue" with K J Marsh and L P Pook, Clarendon Press, Oxford, 1974

P. J. E. FORSYTH

After training as a Mechanical and Electrical engineer with the
Post Office, Peter Forsyth joined the Royal Aircraft Establishment
in 1944. His interest in materials grew from his early investigations of materials and methods of manufacture of captured
German aircraft and rocket components. He started work on
fatigue in 1949, his contributions causing his promotion on
individual Merit to Deputy Chief Scientific Officer in 1977.
His various awards include the Rosenhain Medal of the Institute
of Physics (1967) and the degree of Doctor of Science (NCAA) in
1973.

His main interest has centred round aluminium alloys and the use
of fractography to elucidate the physical mechanisms of fatigue.
His many publications in this field include some heavily referenced
classics.

"Some metallographic observations on the fatigue of metals",
J. Inst. Metals 1951 80 181

"Slip band extrusion and damage", Proc. Roy. Soc. 1957 242 198

"A two stage fatigue fracture mechanism", Proc. Cranfield
Symposium on Fatigue Crack Propagation 1961 1 76

"The Physical Basis of Metal Fatigue", Blackie, 1969

"A unified description of micro and macroscopic fatigue crack
behaviour", Int. J. Fatigue 1983 5 3

G. C. SMITH

Gerry Smith started a shortened war-time metallurgy course at
Cambridge University in October 1942 and after final examinations
in 1944 undertook research work for the Ministry of Supply.
In 1946 he was appointed a Departmental Demonstrator in the
Department of Metallurgy at Cambridge and subsequently a University
Demonstrator then Lecturer.

His research activities have included fatigue, dispersion
hardening, powder metallurgy, ductile fracture and hydrogen-
metal interactions. In addition to his Departmental teaching
he has been a Tutor, Director of Studies and finally Senior
Tutor (1966-1980) of Pembroke College. He has been a member
of the Council of the Institute of Metals and Institution
of Metallurgists, and a member of MOD and Aeronautical Research
Council committees concerned with materials problems.

"Fatigue damage and crack formation in pure aluminium", with
D R Harris, J. Inst. Metals 1959-60 $\underline{88}$ 182

"Changes occurring in the surface of mild steel specimens
during fatigue stressing" with G F Modlen, J. Iron and Steel
Institute 1960 $\underline{194}$ 459

"Crack propagation in high stress fatigue" with C Laird, Phil.
Mag. 1962 $\underline{7}$ 847

"Fatigue strength of sintered iron compacts" with J M Wheatley,
Powder Metallurgy 1963 $\underline{12}$ 141

"Crack closure and surface microcrack thresholds - some experimental
observations" with M N James, Int. J. Fatigue 1983 $\underline{5}$ 75

Thirty Years of Fatigue Crack Growth — an Historical Review

R. A. SMITH

Cambridge University Engineering Department, Cambridge, UK

ABSTRACT

This survey of the last thirty years of fatigue, identifies as the major success the quantification of the rate at which fatigue cracks grow by the use of the alternating stress intensity factor parameter, ΔK. That mechanistic understanding has also advanced is shown by a review of major conferences which have occured in this period. The particular contributions made by P J E Forsyth, N Frost and G C Smith are emphasised where appropriate.

KEYWORDS

Fatigue; fatigue cracks; fatigue crack growth; stress intensity factor; crack growth threshold.

INTRODUCTION

It is convenient to interpret 30 years ago rather liberally, and to take 1953 as a starting point to remind readers just how much the world has changed in under half a lifetime. In 1953 Eisenhower succeeded Truman as the President of the United States, Stalin died and Churchill was Prime Minister of the United Kingdom. The Coronation of Queen Elizabeth II took place in St Paul's Cathedral which was still surrounded by blitz damage. Television broadcasting was in its infancy with black and white pictures being transmitted for four hours per day. Communications were such that America could only take a sound commentary of the Coronation; still pictures were sent across the Atlantic by radio, to be viewed by the public seven minutes after their live creation. There were some 15 working computers in America, compared with 10 in Britain. Food was still rationed in Britain, Everest was climbed, Dennis Compton scored the runs that won the Ashes at the Oval (cricket was still a national sport!). Industry in the UK boomed, with less than 1/4 million unemployed, although the average wage was only £380/year, 40% of cars produced were exported, the ship building industry was greater than that of the USA, Germany and Japan combined and, significantly for our review of the history of fatigue, the Comet had become the world's first passenger jet liner to enter service.

STATUS OF FATIGUE KNOWLEDGE PRIOR TO 1953

The gradual replacement of wooden parts of machines with cast then wrought iron, the invention of steam power and the increasing momentum of the industrial revolution must have caused many unrecognised fatigue failures. It was, however, the rapid rise of railways and failures of, in particular, axles which caused the name 'fatigue' to be coined and experimental investigations to begin in earnest. By the end of the nineteenth century Wöhler's empirical findings of stress range/life relationships and

1

the concept of a fatigue limit were well established. Rules to deal with mean stress effects were proposed and the advice to avoid stress concentrating notches in design was proclaimed, even though quantitative assessments were lacking. However, the mechanisms of fatigue remained a mystery, and fatigue failures were not associated with progressive crack growth, despite, what with hindsight, must have been clear evidence on the fracture surfaces of failed components. It is worth noting that as far back as 1903, Ewing and Humfrey published photomicrographs which clearly showed the progressive development of slip bands on the surface of cyclically stressed iron samples, the slip bands eventually broadening to form cracks. It is indeed strange that this work passed largely unnoticed for more than half a century. By 1948, Nevil Shute, an aircraft engineer, writing in a prophetic novel, made a statement which, even today, would accurately summarise the perception of the wider public of fatigue:

> "Fatigue may be described as a disease of metal. When metals are subjected to an alternating load, after a great many reversals the whole character of the metal may alter, and this change may happen very suddenly. An aluminium alloy which has stood up quite well to many thousands of hours in flight may suddenly become crystalline and break under quite small forces, with most unpleasant consequences to the aeroplane."

THE COMET ACCIDENTS AND SUBSEQUENT ENQUIRY

After two catastrophic accidents involving the total loss of two Comet aircraft, off Elba on 10 January 1954 and, again in the Mediterranean, near Naples on 8 April 1954, the Comet aircraft were grounded and an investigation, which has become a classic of its kind, undertaken. A Comet was taken from service and the fuselage pressure cycled in a hydraulic tank at pressures equivalent to the flight-by-flight loadings. After some 1,830 flights in the tank, together with 1,230 pressurised flights before the test, making a total of 3,060, the cabin structure failed. The starting point of the failure being the corner of one of the cabin windows, see Fig. 1. The Elba accident had occured after 1,290 flights; the Naples accident after 900. In the light of examination of the Elba wreckage, a substantial quantity of which had been recovered from the sea bed, it was deduced that the accidents were caused by structural failure of the pressure cabin brought about by fatigue.

The most likely sources of fatigue initiation were found to be fastener holes located in the highly stressed regions near the corners of the remarkably square cabin windows. Some of the fastener holes were cracked during manufacture! Great difficulties were encountered in estimating accurate stress levels round these fasteners, whilst favourable test results may have been obtained for the detail under consideration because of compressive residual stresses induced by proof testing into the plastic region.

The official report (Cohen et al, 1955) contains much valuable detail, together with some classic (mis-)statements, describing this type of low cycle fatigue failure:

> "the symptoms of failure when it occurs after relatively few cycles (are) therefore less familiar, but they are also less specific. The process has not endured long, and most of the symptoms of a disease which spreads gradually are absent"

(Low cycle fatigue failure may have been thought to be new, but Braithwaite (1854) described the failure of iron girders supporting a beer vat over a century ago. It was some years after the Comet accidents that the Manson/Coffin plastic strain/ endurance work became widely accessible (Manson, 1966). No attempts of crack growth calculations were published; in the light of our knowledge thirty years on, it would have been extremely straightforward to condemn the fatal detail of the design. However, all disasters promote research, and although sadly the Comet accidents were a large nail in the coffin of the British civil aircraft industry, they promoted great interest in fatigue research. Indeed the official report went as far as saying of fatigue that,

> "Although the subject is a difficult one and rapid results cannot be expected, it is fully recognised that the best brains and resources ought to be directed to it."

WORK HARDENING AND FATIGUE IN METALS (1957)

A convenient summary of the studies of fatigue knowledge in 1957, is contained in the reports of a discussion meeting held under the leadership of N F Mott at the Royal Society in February 1957. The title: 'Work-hardening and Fatigue in Metals' served to emphasise the view that fatigue was still thought of as a bulk phenomenon. Indeed in his introduction to the meeting Mott said:

> "It is hardly likely that we shall understand fracture in fatigue without

Comet G-ALYU at Khartoum
shortly after entering
service in 1952

The same Comet in the
pressure testing tank
built for the accident
investigation.

(This and photograph
below from Cohen et al,
1955)

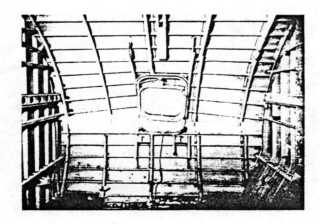

View from inside of
failure produced in the
pressure testing tank.
Note the shape of the
window.

Fig. 1: The Comet accidents and subsequent investigation

first understanding work-hardening and the first few papers (in the meeting) are devoted to this subject"

The metal physicists, whose thoughts were full of the (relatively) new concepts of dislocations, seemed to many at this time to provide a way forward. However, both Smith and Forsyth, Fig. 2, discussed the initial stages of the formation of persistent slip bands and their development into fatigue cracks, albeit in a largely qualitative way, and added detail to the much earlier observations of Ewing and Humfrey. Frost, in a paper co-authored by Phillips, reviewed work undertaken over the previous five years or so, from which an understanding of the propagation phase of the fatigue process was beginning to emerge. Frost noted that more experimental information was needed concerning the extent of the crack tip plastic zone, before the theoretical crack propagation model due to Head (1953), probably the first such model, could be tested. Head had a mechanical model comprising rigid-plastic work-hardening element ahead of the crack tip and elastic elements over the remainder of the infinite sheet, which lead to a relationship of the type

$$\frac{da}{dN} = \frac{C_1 \, \Delta\sigma^3 \, a^{3/2}}{(C_2 - \sigma)\omega_o^{1/2}}$$

with a as crack length, $\Delta\sigma$ stress range and ω_o the size of the crack tip plastic zone. The problem of non-propagating cracks initiated at stress concentration was discussed by Frost and data given on the length v. cycles relationships of such cracks. However, no crack propagation 'laws' were suggested, although this, and associated works, were important milestones in the macroscopic 'engineering' approach to fatigue.

THE CRANFIELD CRACK PROPAGATION SYMPOSIUM (1961)

The changes which occurred in the four years after the meeting just described, are nowhere more evident than in the two volume proceedings of the Cranfield Crack Propagation Symposium held in 1961. It is fair to say that this meeting was a watershed in fracture research and with its passing we entered into the modern age. The range of topics was wide; crack propagation included unstable brittle and ductile fractures as well as sub-critical fatigue crack growth. Although the presentations were firmly focussed on cracks, the engineering community was clearly alarmed by their presence. Thus, Air Marshal Sir Owen Jones in his introduction to the meeting:

"Perhaps our concern might have been indicated less ambiguously if the title of the Conference had been Crack Inhibition Symposium, since we are all much more interested in stopping cracks than in propagating them. However, I am sure I shall be told that it is only when we know what makes cracks propagate that we shall understand how to eliminate or control them."

This meeting further differed from the earlier Royal Society discussion, in that it was an international meeting with over half of the contributions coming from the United States; and it was from the US that the key new ideas emerged. However, two significant papers were presented by Forsyth and Frost, which will be discussed first.

Forsyth's paper 'A Two Stage Process of Fatigue Crack Growth', must rank as one of the most heavily referenced classics of the literature. In it he divided the fatigue process into two stages; Stage I being a crystallographic, shear stress controlled slip process, Fig. 3, occurring to deepen a slip band groove, followed by Stage II growth, normal to the bulk maximum principal tensile stress, giving rise to features on the subsequent fracture surface known as striations. These striations had been reported some two years before the Cranfield meeting (Forsyth and Ryder, 1961), but the original observation was probably due to Zappfe and Worden (1951) who conducted some of the earliest reported fractographs of fatigue fracture surfaces using optical microscopy. However, Zappfe missed the essential point concerning the mode of formation of striations thinking they were caused by some substructural change in the material. Forsyth's Cranfield paper confirmed that one striation corresponded to a single load encursion by illustrating micrographs of programmed loaded specimen fracture surfaces, but he clung to his earlier view (1961), that in the precipitation aluminium alloys of his study, cleavage fracture occurred ahead of the crack tip and that the striation profile was formed by subsequent necking of the intervening material, Fig. 4. This would suggest that striation spacing would be largely governed by the inter-particle distance rather than by stress amplitude and crack length. A year later Laird and Smith (1962), published their findings on high strain growth in aluminium and nickel. They described a model involving sucessive crack blunting on the tensile stroke of the external load cycle followed by re-sharpening during the compression stroke, Fig. 4. This would lead to the peaks of the striations being formed on the compression stroke as opposed to the formation on the tensile stroke in the Forsyth model. In both cases the general form of the surface produced would be the same. Some debate ensued in the succeeding years: it is now acknowledged that the Laird/Smith model is more appropriate for low strength materials, whilst the

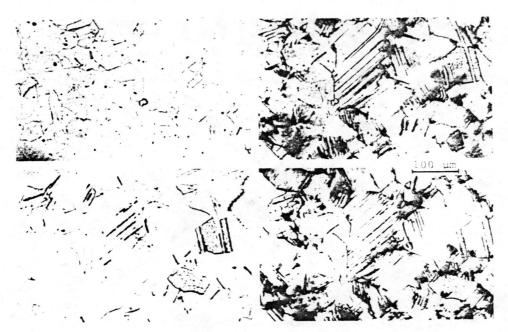

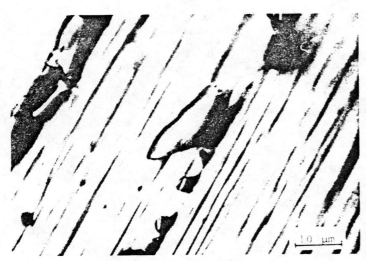

Above: G C Smith's illustration of the development of slip bands into cracks in pure copper.

Left: Extrusion from a slip band in Aluminium/4% copper by P J E Forsyth

Right: A non-propagating crack at a 0.002 in. root radius in L65 Aluminium alloy, N E Frost

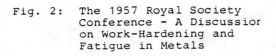

Fig. 2: The 1957 Royal Society Conference - A Discussion on Work-Hardening and Fatigue in Metals

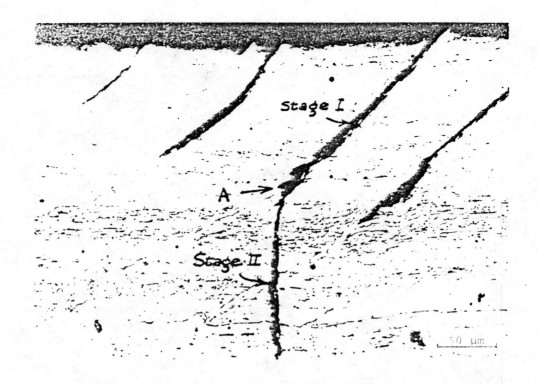

Fig. 3:   The two stage process of fatigue, showing change over from
Stage I (shear controlled) to Stage II (tensile controlled)
growth (Forsyth, 1962).

Forsyth model better fits the observations on higher strength materials where some
cleavage may take place in addition to the plastic blunting and re-sharpening of the
crack tip.   The quantitative aspects of this debate were to resurface when the details
of crack propagation 'laws' were debated.

At the same meeting, Frost, with co-authors Holden and Phillips discussed some
'Experimental Studies into the Behaviour of Fatigue Cracks', undertaken with a view
to determining the critical alternating stress required to propagate a crack of given
length and the laws governing the rate of growth of a growing crack.   These tests
were made on sheet specimens 0.3 inches (7.6 mm) thick, of mild steel, nickel-
chromium alloy steel, copper and a 4½% Cu-Aluminium alloy.   The essential finding was
'Frost's Law', relating the rate of growth of a crack of length a with applied cycles
N, to the alternating stress $\Delta\sigma$ and the crack length, by

$$\frac{da}{dN} = A\Delta\sigma^3 a$$

and that for crack growth to occur, $\Delta\sigma^3 a \geq C$, a critical value, corresponding to the
threshold in modern usage.   These findings were the result of extensive, thorough
and painstaking research over a number of years (e.g. Frost, 1959(a), 1959(b), 1960;
Frost and Dugdale, 1958), which went a great way to simplifying our view of the
mechanics of crack growth.   In particular Frost and Dugdale had noted that the crack
tip plastic zone size increases in proportion to crack length.   However, a more
generally applicable parameter, ΔK, the alternating stress intensity factor, was
aired at the Cranfield meeting by Donaldson and Anderson of the Boeing Company.
This characterisation of fatigue crack growth has become so universal that it is
worth considering its historical origins in some detail.

BRITTLE FRACTURE OF LIBERTY SHIPS

Losses in Allied shipping during the early stages of World War II led to a remarkable
replacement programme in America, based on all welded construction of pre-fabricated
ship sections.   So successful was the management of this scheme, that the later ships
were on the stocks in the builders yards for less than a day!   Unfortunately, many of
the so-called "Liberty" ships suffered from severe brittle fracture problems, some
even separating into two whilst moored in still water, see Figs. 5 and 6.   The
problem proved perplexing; detailed records were kept, vast numbers of impact tests

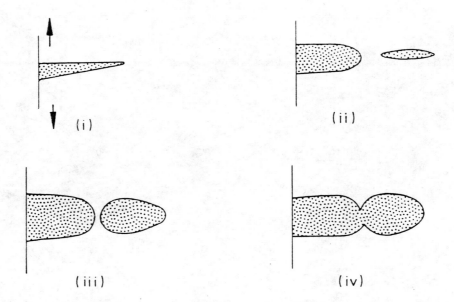

Fracture ahead of the crack tip (i) end of compressive half cycle which may have fractured particles in front of the crack, (ii) tensile half cycle blunts crack and produces void in region of triaxial tension, (iii) thinning of unfractured bridge under biaxial tension, (iv) profile of striation formed. (Forsyth and Ryder, 1961)

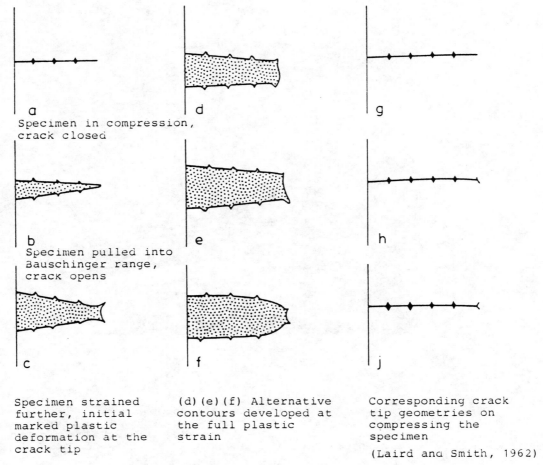

Specimen in compression, crack closed

Specimen pulled into Bauschinger range, crack opens

Specimen strained further, initial marked plastic deformation at the crack tip

(d)(e)(f) Alternative contours developed at the full plastic strain

Corresponding crack tip geometries on compressing the specimen

(Laird and Smith, 1962)

Fig. 4: Alternative models of striation formation

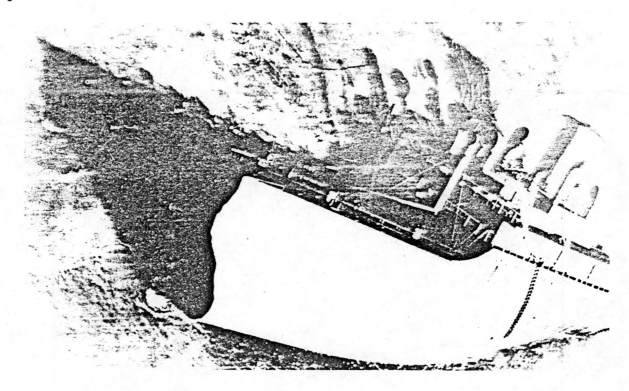

Fig. 5:   Brittle fracture in 'Liberty' ships. Many fractures
          caused complete separation of the ships hull both whilst
          underway or, even more surprising, whilst alongside in
          calm water!

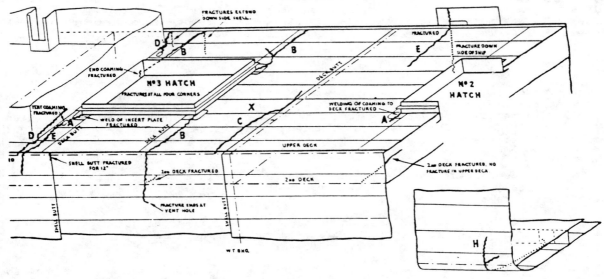

**DIAGRAMMATIC ILLUSTRATION OF FRACTURES
IN WELDED OR MAINLY WELDED SHIPS
DIAGRAM 'A'**

Fig. 6: Locations of cracks at various features of the ships
construction (Brittle Fracture in Mild Steel Plates,
Anon. 1945)

were made and many discussion meetings were held. Typical of these was a meeting
held in Cambridge just after the War (Anon, 1945), to discuss the British and
American experiences. The overemphasis of the metallurgy of the problem is evident,
as is the bafflement of the stars of the scientific community (Baker, Bragg, Orowan,
Mott, G I Taylor) concerning the mechanics of the situation. Later understanding of
the roots of the problem can be summarised as bad geometry of design (sharp corners),
no crack stopping devices in the all welded constructions coupled with occurrences of
low temperatures, poor material and slack quality control of the welding. As far as
fracture theory was concerned, the imposition of near useless impact tests onto the
engineering community, was outweighed by the advantages which stemmed from the concentration over a period of years of the best brains available onto the problems of
cracks in metals.

The theory of stress concentration round elliptical due to Inglis (1913), produced
the classic simple formula for the maximum stress at a notch root:

$$\sigma_{max} = \sigma(1 + 2\sqrt{\frac{D}{\rho}})$$

where σ is the remotely applied nominal stress, D is the notch depth and ρ its root
radius. Whilst this formula proved very useful for predicting stress levels at
relatively blunt notches, a difficulty arose when cracks were considered, since as
$\rho \rightarrow 0$ the maximum local stress $\rightarrow \infty$. Now in ductile materials it was clear that
these high stresses would be limited by yielding of the material, but in both brittle and
ductile materials it was a commonsense practical observation that larger cracks were
more dangerous than smaller ones, a fact which could not be accounted for by the
theory of stress concentrations.

Griffith (1921, 1924) made the important step of diverting attention from local crack
tip conditions by deriving a global instability criteria. His first paper (1921),
"The Phenomena of Rupture of Flow in Solids", starts by stating the origin and object
of the work, which may come as a surprise to many,

"the discovery of the effect of surface treatment - such as filing, grinding
or polishing, on the strength of metallic machine parts subjected to
alternating or repeated loads"

In the event, this paper dealt with the unstable fracture of cracks subjected to monotonic loading, by formulating a necessary thermodynamic criterion for fracture, that the elastic energy available to propagate the crack must be greater than the energy absorbed in creating new crack surfaces. The first paper, although containing interesting detail on experiments with glass, was rather long and discursive, contained an error in the application of Inglis' earlier stress analysis, but nevertheless was an important milestone. The second paper (Griffith, 1924), "The Theory of Rupture", was much more concise. A formula for the remote stress to propagate a central crack in a uniform biaxially loaded sheet was derived:

$$\sigma = \sqrt{\frac{2ET}{\pi a}}$$

with E, the Young's Modulus, a as the semi-crack length and T the surface energy. The theory was applied to inclined cracks in both tensile and compressive stress fields, in the development of which Griffith clearly recognised that a sufficient local condition for fracture must be added to his global necessary condition - that of tensile stresses acting across the prospective crack path (present adherents of "strain-energy density criteria" please note!).

It was the use of surface energy as the only crack tip energy sink that caused the Griffith predictions to be in large error for ductile materials. Griffith's own conclusion was that

> "In the case of plastic crystals, we are further hindered by the fact that rupture is almost invariably preceded by plastic flow, whose nature is still the subject of hot controversy."

It was largely for this reason that Griffith's theory lay unused (and unsung) for nearly twenty years, before the pressing problem of brittle fracture caused it to be re-examined.

THE ORIGINS OF THE STRESS INTENSITY FACTOR

Sneddon (1946) could well claim to have made the first observation that the local stress fields near crack tips were always spatially similar even if the bulk geometries in which the cracks occurred are different. Based on crack tip polar coordinates (r, θ), he observed that for a Griffith type crack, the local stresses could be written as

$$\sigma_{xx} = \sigma \left(\frac{a}{2r}\right)^{\frac{1}{2}} \left(\frac{3}{4} \cos \frac{\theta}{2} + \frac{1}{4} \cos \frac{5\theta}{2}\right)$$

$$\sigma_{yy} = \sigma \left(\frac{a}{2r}\right)^{\frac{1}{2}} \left(\frac{5}{4} \cos \frac{\theta}{2} - \frac{1}{4} \cos \frac{5\theta}{2}\right)$$

$$\tau_{xy} = \sigma \left(\frac{a}{8r}\right)^{\frac{1}{2}} \sin\theta \cos \frac{3\theta}{2}$$

and that for a uniformly stressed penny shaped crack;

> "The most striking feature of the analysis ... is that the expressions for the components of stress in the neighbourhood of the crack differ from those of the two-dimensional case by a numerical factor only."
> [That factor being $2/\pi$.]

He went further by recognising that these local stress fields would still apply if the extent of plastic yielding was small:

> "Thus even for small stresses plastic flow occurs at the corners [tips] of the crack to remove this infinite stress. There is, in fact, no purely <u>elastic</u> solution of the problem; if, however, the internal pressure [stress] is not too large the region of plastic flow will be small and not appreciably affect the distribution of stress at points in the solid at a distance from the corners [tips] of the crack."

The paper goes on to calculate a Griffith criterion for the internal penny shaped crack.

Here the matter lay, until a decade or so later when contributions were made by Irwin (1957a, b) and Williams (1957). Irwin (1957a) presented the local crack tip stresses for the Griffith crack, in the typical form:

$$\sigma_{yy} = \left(\frac{EG}{\pi}\right)^{\frac{1}{2}} \frac{1}{\sqrt{2r}} \cos \frac{\theta}{2} \left(1 + \sin \frac{\theta}{2} \sin \frac{3\theta}{2}\right)$$

Elementary trigonometry reduces the angular distribution to exactly the form of Sneddon's relationship, but Irwin was able to generalise:

> "The stress field near the end of a somewhat brittle tensile fracture, in siutations of generalised plane stress or plane strain, can be approximated by a two-parameter set of equations. The most significant of these parameters, the intensity factor, is $(EG/\pi)^{1/2}$ for plane stress where G is the force tending to cause crack extension."

In the second paper (Irwin, 1957b), the local stresses were written as

$$\sigma_{ij} = K \frac{1}{\sqrt{2r}} f(\theta)$$

the term $K = (2/\pi)(\sigma^2 a)^{1/2}$ being identified as the local scaling factor, the stress intensity factor for a penny shaped crack. (Note that later definitions for SIFs differ by numerical multiple of $\sqrt{\pi}$, i.e.

$$K = \sigma\sqrt{\pi a} \qquad \text{(Griffith crack)}$$

$$K = \frac{2\sigma}{\pi}\sqrt{\pi a} \qquad \text{(Penny shaped crack))}$$

The equivalence;

$$K^2 = EG$$

was derived and finally, the critical value of G at fracture, G_C, was measured for a variety of materials, thus ridding the Griffith criterion of its Achilles heel, surface tension.

[It is worth noting that although the Griffith approach provided the vital stepping stone, for practical purposes it could now be discarded, since a fracture criterion could be based on critical values of local stress controlled by the remote parameter, the SIF. For the Griffith crack the SIF may be considered as the limit,

$$\sigma_{max} \sqrt{\rho} \qquad \text{as } \rho \to 0$$

which on substituting the Inglis relationship and noting $\sqrt{D/\rho} \gg 1$, becomes

$$\sigma \sqrt{\frac{D}{\rho}} \cdot \sqrt{\rho}$$

i.e. $\sigma \sqrt{a}$,

the stress intensity factor, where as the root radius ρ shrinks to zero, the notch depth, D, is equivalent to a crack length, a].

THE USE OF THE STRESS INTENSITY FACTOR TO QUANTIFY FATIGUE CRACK GROWTH

At about the same time that Irwin was laying the foundations of modern fracture mechanics, other workers were seeing if sub-critical crack growth could be treated in a similar manner. Paris (1982), has described these early efforts:

> "In the summer of 1956, Rowe of the Boeing Company asked [Paris] if Irwin's fracture mechanics, at that time based entirely on energy rate analysis could be applied to fatigue cracking. The answer was 'of course not', since the cyclic nature of the fatigue crack tip plasticity could not be accommodated in an energy balance equation! ... Then, Irwin's crack tip stress field analysis (Irwin, 1957a) became available through a preprint in the spring of 1957, and the question was recalled."

However, there were difficulties in obtaining data because of the limitations of testing equipment then available, see Marsh and Smith in this volume. Some two years later (Martin and Sinclair, 1958), tried to correlate growth rates with the G parameter, but erroneously neglected to include the diameter of the central hole with the crack length growing from it in their test specimen, and thus concluded the approach did not work. Paris and co-workers identified this error, found some additional data for the material, and plotted the data as shown on Fig. 7. Paris continues:

> "Ironically, the [resulting] paper was promptly rejected by three leading journals, whose reviewers uniformly felt that 'it is not possible that an elastic parameter such as K can account for the self-evident plasticity effects in correlating fatigue crack growth rates'."

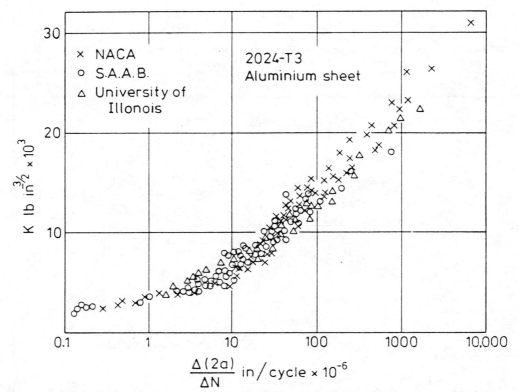

Fig. 7: The original plot with correlated crack growth rates in
2024-T3 Aluminium Alloy from tests conducted by three
independent investigators (Paris et al, 1961)

Thus this major breakthrough was finally published in (with due apologies) the not so readily available 'house journal' of the University of Washington (Paris, Gomez and Anderson, 1961)!

In this original paper, the authors proposed that the growth rate would be a function of the maximum stress intensity factor of the loading cycle and the stress ratio,

i.e. $\quad \dfrac{da}{dN} = f(K_{max}, R) \qquad$ where $R = \dfrac{\sigma_{max}}{\sigma_{min}}$.

Now to return to the Cranfield Symposium of 1961. Donaldson and Anderson made a presentation which introduced this approach to the British members of the audience. Considerable data was produced for aluminium-copper, aluminium-zinc-magnesium, martensitic-low alloy steels and precipitation-hardened stainless steels, still using the K_{max} parameter. A compilation of stress intensity factor solutions for a wide range of geometries was produced, thus finally overcoming the geometry difficulties of the Griffith approach. Paris (1982) has suggested that Anderson was aware in 1962 of the unifying effect of plotting growth rate against $\Delta K/E$, thus identifying the importance of crack tip strains, but no evidence of this approach was presented in Anderson's Cranfield paper. By 1963, however, Paris (1963) was in print co-authored by Erdogan with the use of ΔK "corresponding to stress range" and the classic understatement

"The authors are hesitant but cannot resist the temptation to draw the straight line of slope 1/4 through the data",

leading to the so-called 'Paris' law for the mid-range growth rates,

$$\dfrac{da}{dN} = C(\Delta K)^4$$

see Fig. 8.

In many respects this completed the quantitative framework of fracture mechanics. A great explosion of literature followed over the next two decades, largely adding detail and practical expertise to these simple ideas. The acceptance of the K parameter took perhaps ten years to become almost universal. Previous strong advocates of their own particular 'law' fell by the wayside, e.g. Frost, Pook and Denton (1971), replotted their previous data using fracture mechanics principles and,

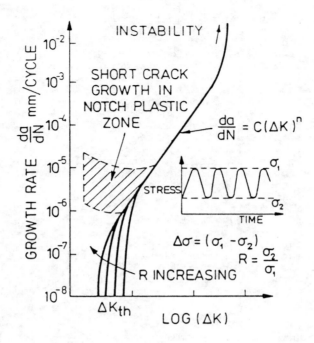

Fig. 8: General trend of fatigue crack growth rates as a function of alternating stress intensity factor. Note the acceleration of "short cracks" in local plastic zones.

somewhat grudgingly, concluded:

"The fatigue crack growth data are presented as master curves of the rate of crack growth vs. the fracture mechanics parameter ΔK, and the method of analysis can be regarded as an alternative to the parameter $\Delta\sigma^3 a$ previously used."

Great debates were held on the value of the index in the 'Paris' law. H W Liu was particularly strong on a second power (e.g. discussion in Paris, 1964), which he argued largely on dimensional grounds. It should also be said that models based on plastic blunting and resharpening also lead to the same conclusion. The situation has now been both clarified and confused with the knowledge that secondary crack advance processes can take place (e.g. burst of cleavage, retardation at inclusions) which upset the simple arguments and lead into the conclusion that, for best accuracy, test data must be obtained from close matching of service conditions.

THE CRACK GROWTH THRESHOLD

The early observation by Frost (1959b), of a relationship between the critical alternating propagation stress and crack length for mild steel, clearly had significant practical consequences, since by using such a parameter, an effective fatigue limit could be defined, even for a cracked component. Many workers produced data for a wide range of materials under different loading conditions, gradually replacing Frost's critical $\sigma^3 a$ parameter, with the equivalent fracture mechanics measure, the fatigue crack growth threshold, ΔK_{TH}. By 1982 (Bäcklund et al, 1982) multi-volumed proceedings of an international conference on this single topic appeared! It rapidly became apparent that the threshold was not quite as simple as it appeared on first sight, the value obtained being strongly dependent on microstructure, crack length, crack closure and stress history. The usefulness of fracture mechanics in the mid-growth rate range, lies, of course, in the swamping of microstructural detail by larger scale continuum mechanics. At slow growth rates down to threshold values, the microstructure can be of scale comparable with the mechanics. Details can be found in the papers by Lindley and Nix, and Beever and Carlsson in this volume. The problem of crack length and the intimately connected closure and stress history effects will be briefly reviewed here.

SHORT FATIGUE CRACKS

It is not surprising that given the near universal acceptance of the K parameter to correlate crack growth for essentially long cracks, attention turned to applying the same ideas to short cracks. Indeed a whole new area of fatigue research has been rediscovered, as the limitations of small scale yielding and continuum behaviour of LEFM (Linear Elastic Fracture Mechanics) have been forgotten and the not so surprising 'anomolous' behaviour of short cracks has been demonstrated. Many reviews of this area have appeared in recent years (e.g. Smith, 1983), and the arguments need not be repeated here. It is sufficient to say that if the crack length is zero, the stress intensity factor is zero and it would be surprising if a parameter of zero value controlled crack initiation. In a similar manner, if the cracks are smaller than microstructural features global fracture mechanics (i.e. applied to the overall structure or specimen) fails (Smith, 1977), and, again for small cracks, to achieve the same nominal ΔK value, such high stresses are required that the crack is no longer contained in an elastic field with plasticity limited to its tip, but is totally embedded in a plastic strain field. These conditions lead to growth rates higher than those predicted by LEFM, see Fig. 8. For the particular cases of cracks growing in the high stress concentration regions of notches, see Smith and Miller (1978), and for cracks initiated by high thermally induced strains, see Marsh (1981). In these regions the links between fracture mechanics crack growth and the low-cycle fatigue concepts developed by Manson and Coffin are evident. It is also worth mentioning that a completely separate methodology has been developed, principally at the University of Illinois by Morrow and co-workers, for use in fatigue estimations where the crack propagation phase occupies a negligible proportion of life. This so called 'local stress-strain' approach (e.g. Osgood (1982)) has its roots in the cyclic bulk properties view of fatigue developed in the 1950s. The principle of the method is to match cyclic stress strain data obtained from small laboratory specimens, to calculated local stresses and strains at the stress concentrating detail of a component at which fatigue is initiated.

CRACK CLOSURE AND LOAD SEQUENCE EFFECTS

It is somewhat surprising that the observation that a fatigue crack propagating under zero-to-tension loading may close at a load greater than zero, appeared many years after crack growth was first studied. Elber (1970) noted that this 'closure' (properly 'non-closure'!) was due to the wake of plastic deformation left behind the tip of a propagating crack and that if the crack was only open for a fraction of the applied range, the crack growth rates should correlate to that part of the ΔK range, ΔK_{eff}, for which the crack was open. Indeed considerable data has been produced to show that ΔK_{eff} can unify fatigue crack growth rates for a range of mean stresses and that growth thresholds are better correlated with ΔK_{eff} than the full range, ΔK. Other causes of closure have been investigated - particularly corrosion products (Suresh et al, 1982) and crack surface morphology (roughness), e.g. Forsyth (1984). Elber (1970) further observed that under variable amplitude loading, crack closure may be responsible for at least part of the interaction effects between stress levels.

It is clear that the plastic deformation behind a crack tip and the plastic zone ahead of the tip, contain information of the loading by which they have been produced. It is therefore optimistic to expect that the current value of ΔK at a particular crack length, contains all the information required to predict the next increment of crack growth. Much work has been reported on the prediction of load sequence effects for single-load excursions (above a constant amplitude value), block loadings of various amplitudes and load signals with increasing degrees of randomness. Prediction techniques range from step-by-step predictions using closure models to representative statistical averages of the applied ΔK signal (e.g. the root-mean-square value of ΔK). Detailed discussion can be found in the paper by Fleck in this volume.

LIVING WITH DEFECTS

Late in 1979 a significant discussion meeting was held at the Royal Society. The title was "Fracture Mechanics in Design and Service - Living with Defects'. Sir Hugh Ford introduced the meeting thus:

> "This is not just another conference on recent advances in fracture mechanics. The theme is the application of the technology already available in the service of safe engineering design and operation. The aim is to review the extent to which both the technology is being applied and the engineer is learning to live with the defects that are inevitably present in his machines and structures."

Compare this with the introductions to the Royal Society meeting of 1957 and the Cranfield Symposium of 1961!

Subcritical crack growth (fatigue, creep and stress corrosion cracking) was reviewed by Tomkins (1981), in a quantitative manner which would have been inconceivable thirty years previously. The extent to which fracture mechanics was being used in industry in both the design and operational support roles was reviewed in contributions from the power generating industries, the petroleum and gas industries (North Sea oil and gas had been developed from nothing in a mere ten years or so!), together with discussions of non-metallics including wood and rubber. Reference was made to one particularly remarkable example from the Central Electricity Generating Board (Neate et al, 1979; Stewart et al, 1977). An examination of a rotor in a 500-megawatt generating set, revealed extensive fatigue cracking, extending over 100 mm from each end of some 24 internal slots. The rotor was 6.5 m long, 1.1 m in diameter, weighed 65 tons and ran at 50 revolutions per second in an atmosphere of hydrogen. No spare rotor was available and manufacture of a replacement could not have been completed in less than two years. The non-availability of the unit for this period would have cost approximately £16 million in replacement-generating costs (efficient large size being replaced by several less efficient smaller units). After a metallurgical examination and fracture analysis, the rotor was pronounced safe for further operation and gave two years' trouble free service before a replacement could be fitted. This is a quite remarkable example of the confidence with which fracture mechanics can now be applied and forms a classic case study of the lessons of the last thirty years including aspects of fatigue crack growth, threshold and mean stress levels, effect of hydrogen on crack growth and use of fractography including striations and beach marks to follow service life.

CONCLUDING REMARKS

The last 30 years have seen a vastly increased activity in fatigue research. The crack propagation phase of fatigue is now well understood. Crack growth rates can be correlated using the engineering based alternating stress intensity parameter. Much current research is filling in of secondary detail. This area of crack growth is now well enough understood to have been utilised to good effect in many industries.

The crack initiation phase of fatigue, not reviewed in this paper, is well understood qualitatively, but not quantitatively.

More generally, it is interesting to note that key advances in this topic have occurred in an almost accidental manner rather than as the fruits of directed research, although resources have been focussed by well publicised failures.

REFERENCES

Anon. (1945). Brittle Fracture in Mild Steel Plates, Proc. Conf. Cambridge University, Oct. 26, 1945, The British Iron and Steel Research Association

Bäcklund, J, A F Blom and C J Beevers (Eds.) (1982). Fatigue Thresholds - Fundamentals and Applications, EMAS, Warley, Birmingham

Braithwaite, F (1854). On the fatigue and consequent fracture of metals. Proc. Inst. Civil Engrs., 13, 463-475

Cohen, Baron L of Walmer, W S Farren, W J Duncan and A H Wheeler (1955). Report of the Court of Inquiry into the Accidents to Comet G-ALYP on 10 January, 1954 and Comet G-ALYY on 8 April 1954, HMSO, London

Donaldson, D R and W E Anderson (1962). Crack propagation behaviour of some aircraft materials. In Proc. Crack Propagation Symposium, Vol. II, Cranfield College of Aeronautics, pp. 375-441

Ewing, J A and J C W Humphrey (1903). The fracture of metals under repeated alternating stresses. Proc. Roy. Soc., Series A, 200, 241-250

Elber, W (1970). Fatigue crack closure under cyclic tension. Eng. Fract. Mech., 2, 37-45

Forsyth, P J E (1957). Slip-band damage and extrusion. Proc. Roy. Soc., Series A, 242, 198-202

Forsyth, P J E (1962). A two stage process of fatigue crack growth. In Proc. Crack Propagation Symposium, Vol. I, Cranfield College of Aeronautics, pp. 76-94

Forsyth, P J E (1984). Fatigue crack closure and its dependence on fracture features. J. Mats. Sci., 19, 1836-1850

Forsyth. P J E and D A Ryder (1961). Some results of the examination of aluminium alloy specimen fracture surfaces. Metallurgia, 63, 117-124

Frost, N E (1959a). Propagation of fatigue cracks in various sheet materials. J. Mech. Eng. Sci., 1, 151-170

Frost, N E (1959b). A relation between the critical alternating propagation stress and crack length for mild steel. Proc. Instn. Mech. Engrs., 173, 811-836

Frost, N E (1960). Notch effects and the critical alternating stress required to propagate a crack in an aluminium alloy subject to fatigue loading. J. Mech. Eng. Sci., 2, 109-119

Frost, N E and D S Dugdale (1958). The propagation of fatigue cracks in sheet specimens. J. Mech. and Phys. Solids, 6, 92-110

Frost, N E, J Holden and C E Phillips (1962). Experimental studies into the behaviour of fatigue cracks. In Proc. Crack Propagation Symposium, Vol. I, Cranfield College of Aeronautics, pp. 166-187

Frost, N E and C E Phillips (1957). Some observations on the spread of fatigue cracks. Proc. Roy. Soc., Series A, 242, 216-222

Frost, N E, L P Pook and K Denton (1971). A fracture mechanics analysis of fatigue crack growth data for various materials. Eng. Fract. Mech., 3, 109-126

Griffith, A A (1921). The phenomena of rupture and flow in solids. Proc. Roy. Soc., Series A, 221, 163-198

Griffith, A A (1924). The theory of rupture. In Proc. 1st Int. Congress. Appl. Mech., Delft, pp 55-63

Head, A K (1953). The growth of fatigue cracks. Phil. Mag., 44, 925-38

Inglis, C E (1913). Stresses in a plate due to the presence of cracks and sharp corners. Trans. Inst. Naval Arch., LV, 219-230 (and discussion pp. 231-241)

Irwin, G R (1957a). Relation of stresses near a crack in the crack extension force. In Proc. 9th Int. Congress Appl. Mech., Brussels, Vol. 8, pp 245-251

Irwin, G R (1957b). Analysis of stresses and strains near the end of a crack traversing a plate. J. Appl. Mech., 24, 361-364

Laird, C and G C Smith (1962). Crack propagation in high stress fatigue. Phil. Mag, 7, 847-857

Mason, S S (1966). Thermal Stress and Low-Cycle Fatigue, McGraw-Hill, New York

Marsh, D J (1981). A thermal shock fatigue study of Type 304 and 316 stainless steels. Fat. Eng. Mats. and Struct., 4, 179-195

Martin, D E and G M Sinclair (1958). Crack propagation under repeated loading. In Proc. Third US COngress Appl. Mech., 595-604

Neate, G J, G M Sparkes, A T Stewart and H D Williams (1979). Application of fracture mechanics to industrial problems. In R A Smith (Ed.), Fracture Mechanics, Current Status, Future Prospects, Pergamon, Oxford, 69-90

Osgood, C C (1982). Fatigue Design, 2nd Edn., Pergamon, Oxford, pp. 74-86

Paris, P C (1964). The fracture mechanics approach to fatigue. In J J Burke, N E Reed and V Weiss (Eds.) Fatigue - An Interdisciplinary Approach, Syracuse University Press, pp. 107-132

Paris, P C (1982). Twenty years of reflection on questions involving fatigue crack growth. In J Bäcklund, A F Blom and C J Beevers (Eds.) Fatigue Thresholds - Fundamentals and Applications, EMAS, Warley, Birmingham, pp. 3-10

Paris, P C and F Erdogan (1963). Critical analysis of crack propagation laws. Trans. ASME, J. Basic Eng., 85, 528-534

Paris, P C, M P Gomez and W E Anderson (1961). A rational analytic theory of fatigue. The Trend in Engineering, 13, 9-14

Shute, N (1948). No Highway, Heinemann (Many subsequent editions)

Sneddon, I N (1946). The distribution of stress in the neighbourhood of a crack in an elastic solid. Proc. Roy. Soc., Series A, 187, 229-260

Smith, G C (1957). The initial fatigue crack. Proc. Roy. Soc., Series A, 242, 189-197

Smith, R A (1977). On the short crack limitations of fracture mechanics. Int. J. Fract., 13, 717-720

Smith, R A (1983). Short fatigue cracks. In J Lankford, D L Davidson, W L Morris and R P Wai (Eds.) Fatigue Mechanisms: Advanced in Quantitative Measurement of Physical Damage, ASTM STP811, pp. 264-279

Smith, R A and K J Miller (1978). Prediction of fatigue regimes in notched components. Int. J. Mech. Sci., 20, 201-206

Stewart, A T, K A Haines and H D Williams (1977). Influence of residual stress on crack propagation in an alternator rotor, In P Stanley (Ed.) Fracture Mechanics in Engineering Practice, Applied Science, London, pp. 323-338

Suresh, S, D M Parks and R O Ritchie (1982). Crack tip oxide formation and its influence on fatigue thresholds. In J Bäcklund, A F Blom and C J Beevers (Eds.) Fatigue Thresholds - Fundamentals and Applications, EMAS, Warley, Birmingham, pp. 391-408

Tomkins, B (1981). Subcritical crack growth: fatigue, creep and stress corrosion cracking. Proc. Roy. Soc., Series A, 299, 31-44

Williams, M L (1957). On the stress distribution at the base of a stationary crack. Trans. ASME, J. App. Mech., 24, 109-114

Zappfe, C A and C O Worden (1951). Fractographic registrations of fatigue. Trans. Amer. Soc. Metals, 43, 959-969

Thirty Years of Fatigue Testing Equipment

K. J. MARSH★ and R. A. SMITH★★

★National Engineering Laboratory, Glasgow, UK
★★Cambridge University Engineering Department, Cambridge, UK

ABSTRACT

The fatigue testing equipment available in the mid-fifties, particularly that appropriate to crack propagation tests, is reviewed, together with early approaches to such testing. Several major developments are then discussed. Firstly, the advent of servo-hydraulic testing machines with their greatly increased flexibility is described. This also had profound effects on the parallel field of structural fatigue testing and some illustration of this activity at NEL is given. Secondly, the advent of digital computers, and their rapid proliferation as on-line adjuncts to fatigue testing is covered. In these days when the ubiquitous 'micro' permeates most endeavours it is difficult to contemplate signal generation, data analysis and control systems without computers, another revolution in the last 30 years. Finally, some mention is made of the development of equipment for crack detection in components and of the range of available techniques for laboratory crack measurement.

KEYWORDS

Fatigue; testing equipment; crack propagation; servo-hydraulic equipment; computer-aided testing; structural testing; crack detection; crack measurement techniques.

INTRODUCTION

Although this Conference deals with fatigue-crack growth, it is the intention in this paper to treat the development of fatigue testing equipment a little more broadly, since such developments have applied simultaneously to many aspects of fatigue testing.

It is instructive to consider just what fatigue testing equipment was generally available 30 years ago, in the mid-fifties, and, in today's terms, how simple this was. At that time, enormous quantities of fatigue data, in the form of stress amplitude versus life (S/N) curves or fatigue limit determinations, had been generated using extremely simple rotating-bending machines. Figure 1 shows a classic machine, a derivative of a type developed by Wöhler for his studies on railway axle steels more than a century ago; such machines were certainly still operating in the mid-sixties. It consists simply of a specimen clamped in a rotating chuck, to which a constant bending moment over the gauge section was applied by a simple dead-weight system. This elegantly simple machine was unfortunately not very useful for studying crack propagation, although at the National Engineering Laboratory (NEL) the use of very sharply-notched specimens did allow some study of the situation where a crack initiated very quickly and most of the life was spent in crack propagation (Frost, 1960). Indeed, Fig. 2 shows block-programme loading markings on the fracture face of a sharply-notched rotating-bending specimen tested in a machine very little more sophisticated than that described.

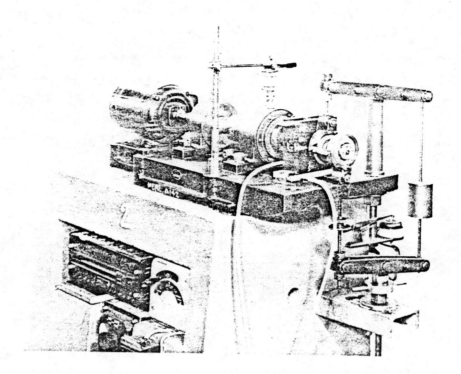

Fig. 1: Simple dead-weight rotating bending fatigue machine

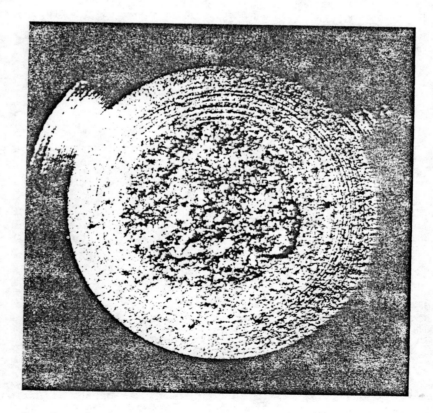

Fig. 2: Programme markings on a sharply-notched rotating-bending
specimen (Marsh, 1968)

However, it was obvious that serious crack propagation studies required the use of direct-stress or 'push-pull' machines. Those available 30 years ago were also very simple machines and fell into two categories, 'resonant' machines and 'brute-force' machines. In 'resonant' machines, the specimen formed part of a vibrating spring-mass system excited to resonance. The best known of these were the Amsler Vibrophore machines, which used electromagnetic coil excitation and could operate at frequencies from about 50 - 300 Hz, and the Schenck Pulsers (Fig. 3) where the spring-mass system was excited by an out-of-balance rotating mass and the machine operated on the rising slope of the resonance curve, higher loads corresponding to higher frequencies. An example of a 'brute-force' machine was the pulsating hydraulic pressure machine, such as the Losenhausen range (Fig. 4), still in use today in many laboratories.

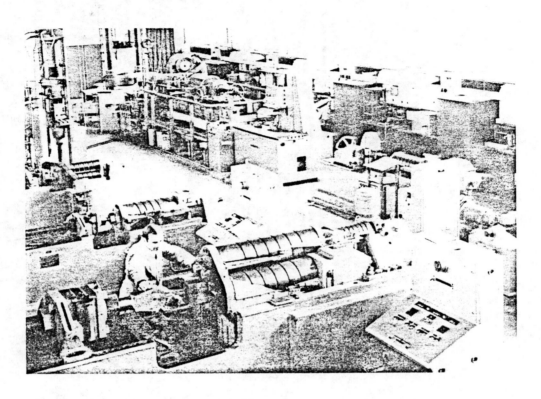

Fig. 3: Schenck Pulser 600 kN load capacity machine

Specimens, too, tended to be very simple. Much of Frost's early work (eg Frost, 1959) was carried out on 250 mm wide sheet specimens (Fig. 4), approximately 2.5 mm thick, with a central slit sawn from a small hole, the ends being sharpened with a serrated razor blade. Readings of half-crack length were taken at each end of the crack and on each side of the sheet (four readings) either directly by eye, or with the aid of a small low-powered hand-held telescope. Crack-growth curves, for a given mean and alternating stress, were plotted manually, crack length versus number of cycles; crack-growth rate was also determined manually and subjectively by drawing tangents to these curves, to allow relationships between growth rate and various empirical parameters to be deduced. This was, of course, long before the advent of fracture mechanics and the ubiquitous stress intensity factor; many of Frost's classic papers were published before 1960.

This, then, was the fatigue testing equipment available 30 years ago. The surprising fact is how little change there was in the type of equipment used for almost half of the period between then and now. For example, in the late 1960s work was published by Marsh and Mackinnon (1968) on random- and block-loading cumulative damage tests using simple resonant machines and crude test rigs based on simple vibrators. However, the second half of this 30-year period did bring radical changes in fatigue testing equipment which are covered in the subsequent sections.

Fig. 4: Losenhausen UHS40 pulsating pressure fatigue machine,
with 250 mm wide sheet specimen (Frost, 1959)

SERVO-HYDRAULIC TESTING EQUIPMENT

The first major and perhaps most dramatic development in fatigue testing equipment
was the advent and rapid proliferation of the servo-hydraulic testing system. Here
the brute-force of hydraulic power was harnessed to the sophistication of modern
electronic control systems. The operation of a hydraulic load-applying ram is controlled by a high-response servo-valve responding to the differential feed-back from
an appropriate sensor (load cell, deflection transducer, strain gauge) in a closed-
loop control system. It is not over-emphasis to state that this innovation revolutionised fatigue testing. It came at a time when the limitations of constant-
amplitude sinusoidal load variations had been realised and crude mechanical block-
programme systems, or electromagnetic vibrator based random-loading rigs were being
applied to cumulative damage studies. Hence the flexibility of the servo-hydraulic
system, with which any waveform within the positive and negative load or deflection
limits of the machine could be applied, was of paramount importance. Inputs could
be provided from function generators, from magnetic tape or, subsequently, from
digitally-generated signals. The way was therefore open to applying service-recorded
load histories or other derived randomly-varying inputs rather than crude simulations
of these. The control loop could be closed in load, deflection, or strain control
modes. Large load capacities could be readily provided given sufficient hydraulic
power and, similarly, although less important for crack propagation testing, large
deflections could be applied compared to conventional mechanical machines.

There were some disadvantages of course. Power requirements were massively greater
than those of resonant machines, particularly if high frequencies and large load or
stroke capacity were required. As early as 1966 the merits of resonant compared to
servo-hydraulic machines were discussed by Haas and Kreiskorte. Servo-hydraulic

Fig. 5: Servo-hydraulic machine of ± 600 kN load capacity

machines began to be installed in UK laboratories in the middle or late 1960s, for example the first servo-hydraulic system was installed at NEL in 1968, approximately. Figure 5 shows a typical machine consisting of a load frame, a controller and the loading actuator generally mounted in the base. In some cases it is convenient to have the actuator mounted in the cross-head as in the ± 2.5 MN machine shown in Fig. 6; this allows the possibility of also testing structures in the machine as is shown in Fig. 7. Such equipment is commonplace now; however in the late 1960s it was novel to the fatigue workers pioneering its use.

At this stage it is possibly appropriate to consider the other major impact of the advent of servo-hydraulic systems, namely the use of loading actuators to apply simulations of service load histories to engineering structures, part-structures or components, often involving several discrete load-inputs on the same structure to obtain adequate simulations. Since most such structures were welded fabrications, this type of testing was, in a very real sense, crack propagation testing, but it was only possible because of the availability of servo-hydraulic equipment. From the late 1960s NEL has concentrated intensively on this aspect of fatigue testing and much work was done on testing vehicle structures and on simulating firing loads in prototype artillery gun structures (Marsh, 1976).

Figure 8 shows a vehicle cab structure undergoing fatigue testing on a 'road-simulator' servo-hydraulic test rig and Fig. 9 fatigue cracking in the structure. Figure 10 shows the 105 mm Light Gun developed by RARDE. Extensive prototype development of the welded high-strength steel supporting structure was carried out at NEL, simulating programmes of firing loads of different severity and corresponding to different angles of elevation of the armament, by means of a two-actuator vectored-loading system as shown in Fig. 11. Figure 12 shows the propagation of a crack in the structure. The 1k markings correspond to one thousand simulated firings. All this occurred during the early 1970s and was perhaps typical of the drive to find new applications for servo-hydraulic equipment which existed then.

22

Fig. 6: Servo-hydraulic machine of ± 2.5 MN load capacity, with
         loading actuator in the cross-head

Fig. 7: The same machine set up for structural testing

Fig. 8:   Vehicle cab fatigue test on the NEL Road Simulator

Fig. 9:   Three-inch long fatigue cracks in the vehicle cab from
the corner of the window cab

Fig. 10:   A modern 105 mm light artillery gun

It is perhaps instructive to go even further back into the early use of servo-hydraulic equipment; a historial curiosity was the first structural fatigue test carried out at NEL in 1969. As Fig. 13 shows, it was necessary to carry out a service-loading simulation test on a prototype telescopic crane jib. The only two actuators installed at NEL at that time were of short stroke, although ample load capacity. To achieve the necessary deflection in the floppy crane jib, the two actuators were fixed in series to double the stroke, but had then to be operated in open-loop control mode. It was a crude first service-loading simulation structural fatigue test, but much was learnt from the exercise (Marsh and Morrison, 1970).

It would be possible to give many more examples. However, the point is essentially made. All aspects of fatigue testing, including crack propagation testing, were revolutionised by the advent of the servo-hydraulic testing system.

THE DIGITAL COMPUTER

The second major development to affect fatigue testing was the advent of the digital computer. Thirty years ago, in the mid-1950s, the computer played no significant part in fatigue. On the analysis side, most calculations were limited to those which could be readily handled with a slide-rule. Input devices to testing equipment, when not some mechanical form of simple-harmonic-motion generator, were very simple electronic waveform generators. By the 1960s this had begun to change.

It is difficult to put a precise chronology on the process, but it is instructive to attempt to make an assessment, however subjective. By the mid-1960s off-line fatigue data analysis, using central mainframe facilities, was becoming common. However, it was not until about 1970 that the availability of relatively small, relatively cheaper mini-computers began to allow their use on-line, for signal

Fig. 11: Two-actuator firing load simulation test on the Light Gun structure

Fig. 12: Progression of cracks at a weld in the Light Gun structure

Fig. 13: Early structural fatigue test on a telescopic crane jib

generation and data analysis purposes. The need to generate random-type signals either to a prescribed power spectral density or to simulate recorded randomly-varying service-loading histories, and the limitations of analogue random signal generators led to rapid developments in digitally-generated signals to provide the inputs to the flexible servo-hydralic loading systems.

At about the same time, i.e. during the early 1970s, the on-line data monitoring/analysis techniques were also developing rapidly and obviously signal generation and monitoring necessarily go together to some extent. The computer might be used, for example, to monitor the mean and r.m.s. values of random signals, comparing output with demand, from a structural fatigue test, or to carry out running fatigue damage calculations involving peak counting methods such as rain flow analysis for a uni-axial testpiece. By the mid-1970s, in uniaxial loading systems, well established computer control systems had been developed, for example for low-cycle constant strain amplitude testing, whereby computer input generation was a response to feed-back from digitised strain output signals.

It is interesting to note that in an international conference held in 1976 (Bathgate, 1976), rather less than half of the papers involved the use of computers in fatigue testing, whether for signal generation, monitoring and analysis or uniaxial computer control. By 1978 it was possible to run an international conference (Sherratt, 1978) entirely on the topic of the application of computers in fatigue, although only one paper dealt specifically with computerised crack propagation testing and that restricted to random signal generation and on-line analysis of crack-growth data. It seems to have been somewhat later, at the beginning of this decade, that full computer control was applied to crack propagation testing, allowing, for example, constant ΔK tests, automatically reducing load inputs on the basis of feed-back from potential drop crack-growth data.

In multi-channel service-loading structural testing, it had become obvious that the response of an actuator in a closed-loop system to its nominal demand signal could be affected by the way the structure responded to other simultaneous inputs. This 'interaction' problem could only be handled by sophisticated involvement of the computer in the control loops, analysing and modifying responses. Such systems have also become commercially available in the last few years.

In general, the advent of micro-electronics, and the exponential increases in power and decreases in cost of microprocessors must surely mean that there is virtually no limit, other than in software ingenuity, to the applications of digital techniques in sophisticated fatigue testing.

CRACK DETECTION IN COMPONENTS

For completeness, it is worth while to consider very briefly the development of equipment for the detection of cracks in service components (laboratory techniques for specimens are considered at greater length in the next section).

The main techniques available today were established over thirty years ago; dye penetrants, magnetic flux, x-rays and ultrasonics. They were then employed in crude forms for the detection of inherent defects in castings and fabrications. The only method which has been significantly refined over the past thirty years is that of ultrasonics, but with this technique the real pay-off has been in medical rather than engineering applications. Some minor techniques have been developed; the resonance listened to by the railway wheel tapper has now developed as the carefully monitored vibration signature of large generator rotors. By the late 1970s both acoustic emission and alternating current potential difference (a.c.p.d) were being applied to larger structures such as tubular steel welded joints in the UK Offshore Steels Research Project, albeit under laboratory, rather than service, conditions. Figure 14 shows the largest nodes involved, of 1.8 m chord diameter, 75 mm thick, while Fig. 15 shows a typical fatigue crack around the toe of a weld.

Since dye penetrants and magnetic flux methods are only suitable for surface-breaking cracks, and the x-ray method depends on differential absorption, thus being better suited to voids rather than cracks, the ultrasonic method remains the most useful technique for detecting interior cracks. It is probably true to say that developments in this area have lagged far behind our improved quantitative capability caused by the rise of 'fracture-mechanics'. The minimum crack size which <u>may</u> be detected in good conditions is probably about 1 mm; in poor service conditions, cracks of some tens of millimetres in length may go undetected. In many situations, fatigue cracks of these sizes will only be present when the majority of the fatigue life has been consumed; thus, rapid developments in crack detection and sizing are urgently needed if we are to reap the full practical benefits offered by our increased theoretical knowledge.

Fig. 14:   1.8 m diameter tubular welded joints testing in the UK Offshore Steels Research Project

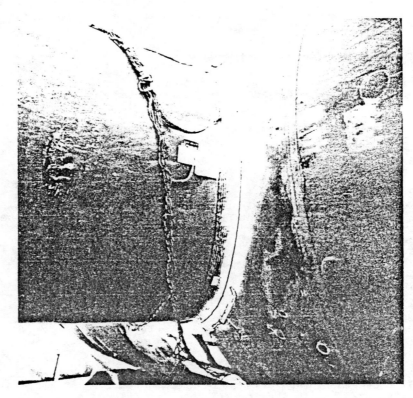

Fig. 15:   Fatigue crack at the toe of the weld in a tubular joint

GENERATION OF FATIGUE CRACK PROPAGATION DATA

Although stress intensity factor calibrations are available for a wide range of geometries (e.g. Tada, Paris and Irwin, 1973; Rooke and Cartwright, 1976), the vast bulk of data generation is from standard specimens, using a procedure suggested by ASTM (1982).   A wide range of laboratory crack measurement techniques is available, which can be divided into direct and indirect methods - the indirect methods require some form of prior calibration, but since these techniques are the ones which can provide a feedback signal into the control loop of a servo-hydraulic testing machine, they have become highly developed.   Figure 16 summarises the most popular measurement techniques.

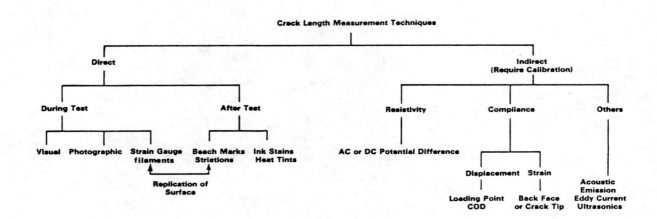

Fig. 16:   Various laboratory crack measurement methods

A very timely review of the state of the art took place at the symposium on the measurement of crack length and shape during fracture and fatigue, held at the University of Birmingham in 1979. In particular, the guidelines to the selection of techniques, in the proceedings of that meeting (Beevers, 1980), provide useful information for the relatively uninitiated. The subsequent volume (Beevers, 1982) explored more recent techniques and particularly treats the move towards automated testing.

The measurement of the fatigue crack propagation threshold requires the use of many of the above techniques because particular care must be taken over the stress history of the specimen prior to the threshold being reached (from a higher stress intensity value). This area of testing is still very immature and despite a large volume of work, e.g. a recent conference of over 70 papers, no agreed standard test procedure exists, although a method is currently being discussed by ASTM (Bäcklund, Blom and Beevers, 1982).

Finally, it is worth noting that considerable efforts are being made to apply the principles of fracture mechanics to small fatigue cracks, often in the range of one millimetre to tens of microns. Progress in this so-called 'short-crack problem', is hampered by the difficulties of crack length and shape measurement at these small size scales.

CONCLUDING REMARKS

The development of fatigue testing equipment has clearly played a major part in the way advances in fatigue crack propagation knowledge have developed in the last 30 years. In general, it would seem that there was very little change in equipment over the first 15 years of this period. During the next 10 years there was a step change with the advent of servo-hydraulic testing equipment and of on-line computers, and the rapid proliferation of both techniques. However, the last five years and the advent of micro-electronics have seen an even more rapid increase in the development of techniques.

Although standardised techniques have emerged for the production of fatigue crack propagation data on laboratory specimens, the improvements in testing techniques have made the application of service loadings to whole structures or prototypes a desirable method of fatigue assessment. No major improvements have been made in our ability to detect and measure cracks in real service environments and this lack of practical NDT techniques for cracks less than about 1 mm in size, remains a serious obstacle, hindering the practical application of fracture mechanics.

ACKNOWLEDGEMENTS

This paper is published by permission of the Director, National Engineering Laboratory, Department of Trade and Industry. It is Crown copyright. We would like to acknowledge the help of numerous members of Structural Testing Division, NEL, in assembling this review.

REFERENCES

ASTM (1982). Annual book of ASTM standards. Part 10, E647-81, 772-790

Bathgate, R G (Ed.) (1976). Proc. Internat. Conf. on Fatigue Testing and Design. City University, London, Society of Environmental Engineers Fatigue Group

Beevers, C J (Ed.) (1980). The measurement of crack length and shape during fracture and fatigue. Engineering Materials Advisory Services Ltd.

Beevers, C J (Ed.) (1982). Advances in crack length measurement. Engineering Materials Advisory Services Ltd.

Bäcklund, J, Blom, A F, and Beevers, C J (Eds) (1982). Proc. Internat. Symp. on Fatigue Thresholds, Stockholm. Engineering Materials Advisory Services Ltd.

Frost, N E (1959). Propagation of fatigue cracks in various sheet materials. J. Mech. Engng. Sci., 1, 151-170

Frost, N E (1960). Notch effects and the critical alternating stress required to propagate a crack in an aluminium alloy subject to fatigue loading. J. Mech. Engng. Sci., 2, 109-119

Haas, T, and Kreiskorte, H (1965-66). Critical comparison of modern fatigue testing machines with regard to requirements and design. Proc. Instn. Mech. Engrs., 180(3A), 31-45

Marsh, K J, and Mackinnon, J A (1968). Random-loading and block-loading fatigue tests on sharply notched mild steel specimens. J. Mech. Engng. Sci., 10, 48-58

Marsh, K J (1968). Further varying-amplitude fatigue tests on sharply notched mild steel specimens. NEL Report No. 373. East Kilbride, Glasgow: National Engineering Laboratory

30

Marsh, K J and Morrison, A J (1970). Fatigue testing of large engineering components and structures. NEL Report No. 466. East Kilbride, Glasgow: National Engineering Laboratory

Marsh, K J (1976). Full scale testing of components and structures, 22.1-22.5, in Bathgate (1976)

Rooke, D P and Cartwright, D J (1976). Compendium of stress intensity factors, HMSO, London

Sherratt, F (Ed.) (1978). Proc. Internat. Conf. on Applications of Computers in Fatigue. Warwick University. Society of Environmental Engineerings Fatigue Group

Tada, H, Paris, P and Irwin, G (1973). The Stress Analysis of Cracks Handbook, Del Research Corp., Hellertown, Pa, USA

Models of Fatigue Crack Growth

J. F. KNOTT

Cambridge University Metallurgy and Materials Science Department,
Cambridge, UK

ABSTRACT

This chapter examines some of the various expressions which have been used to
characterise the growth of fatigue cracks subjected to alternating applied stress or
strain and attempts to reconcile these in terms of the micro-mechanisms of crack
growth. This field is one to which the work of Norman Frost, Peter Forsyth and
Gerry Smith has made large contributions. The main topic treated is growth under
quasi-linear-elastic (LEFM) conditions and, here, attention is paid to the range of
crack-tip opening-displacement ($\Delta\delta$) produced in each loading cycle. This can be
related, on the one hand, to micro-mechanisms of growth, involving reversed slip at
the crack tip: on the other, to applied values of stress-intensity range (ΔK) provided that "closure" effects are not significant. In general, a model based on $\Delta\delta$
is able to characterise growth in Regime B (where the "Paris law" holds) and predicts some mild effects of stress-ratio (R) on growth rate, in agreement with
experimental observations. At high values of maximum stress intensity (K_{max})
"static" fracture modes become significant, and stronger effects of stress-ratio are
observed. Appropriate modelling needs to take separate account of components
dependent on ΔK and K_{max}. In the near-threshold region, effects of stress-ratio are
predicted by the $\Delta\delta$ model, but "closure" effects can alter the relationship between
the crack-tip value of $\Delta\delta$ and the applied stress intensity ΔK. Contributions to
closure occasioned by oxide growth, surface roughness and the distribution of plasticity in the "wake" behind the crack-tip are discussed. At low values of ΔK, in
particular, growth-rates are very much slower in vacuum than in air. This is attributed to the extent to which crack-tip plastic flow is geometrically reversible and,
in this sense, affects the proportionality between crack growth-rate, (da/dN) and $\Delta\delta$.

KEYWORDS

Fatigue, fatigue-crack growth-rate, linear-elastic fracture mechanics, fatigue
thresholds, "static" fracture modes, closure.

INTRODUCTION

The engineering importance of fatigue-crack growth can scarcely be overemphasised.
An extremely large number of structures and components, particularly in the critical
areas of transportation and energy-production, are fatigue-limited in design. To
specify the interaction between fatigue-life, applied cyclic stress and initial
defect size, it is necessary to possess a detailed knowledge of crack growth-rates
in service conditions. Usually, the cyclic stresses are well below yield, so that
the problem can be treated by linear elastic fracture mechanics (LEFM), but account
must be taken of stress-concentrations; mean stress, whether induced by the applied
stress-cycle or by residual stress; variable-amplitude loading; and effects of

environment.

Our knowledge of the processes which cause decohesion and crack advance in fatigue is incomplete. It is appropriate to quote the views of Sir Alan Cottrell (1977) when interviewed on behalf of the Fourth International Congress on Fracture (ICF4).

ICF4:   "How much further do you think we ought to go in pursuing the theory of fracture?"

Sir Alan: "It depends what you want to do. I think there are some areas of the subject which are not understood at all: a lot of fatigue and fractures of that kind. We understand some aspects of fatigue fracture now but many of the others are not understood at all, in terms of the atomic processes going on" . . . . . .
          . . . . . . "Well, again, I would say that I think the disappointments have been that we still don't understand a great deal about fatigue . . . . I think this is because these are processes at the atomic level where you've really got to see what is going on and we still haven't got instruments that will quite allow us to do that fully". . . . . there are problems in the work-hardening stage, because we don't really understand why there is the localization in specific active slip planes. Also, the oscillating mechanical system is geometrically reversible . . . and ideally, all the atoms should go back again. I'd say that there is some second-order effect which is opening up a crack in the surface of these active slip bands and that is something else that we don't understand. Maybe we know the answer already, but haven't proved it. It may be true that gaseous adsorption makes it structurally irreversible but we don't have a hard proof of that. It is just a good surmise."

In general, the state of fundamental understanding of fatigue processes today is little better than it was at the time of the interview, although detailed studies of the fatigue-crack "threshold" (Ritchie 1981, Suresh, Zamiski and Ritchie 1981, Beevers 1983, 1985) have enabled a better, physical appreciation to be gained of the factors which influence crack arrest. There are many models of fatigue-crack growth (e.g. Paris and Erdogan 1963, Liú 1961, Weertman 1969, Lardner 1967, Tomkins 1968, McEvily 1977) but few of these are able to incorporate the full range of experimental observations. No physically-based model is at present able to predict quantitatively the crack growth increment per cycle for a given applied stress-spectrum and environment.

The present paper attempts to highlight the main experimental findings which must be incorporated into any realistic model, examines the ways in which this might be done, and then comments on the extent to which experimental results have been taken into account in existing theoretical treatments. The discussion centres primarily on LEFM conditions, although observations made on initially smooth specimens in high-strain fatigue are also examined, because these are of crucial importance with respect to a physical appreciation of the processes of fatigue-crack growth. Throughout, the importance of the roles played by Norman Frost, Peter Forsyth and Gerry Smith will be clear. Frost was one of the first to relate crack growth rate to the parameters of applied stress-amplitude and instantaneous crack length (Frost, Holden and Phillips 1961) which are an integral part of any fracture mechanics treatment. Forsyth and Smith both made detailed observations of fatigue initiation and propagation but Forsyth is probably more associated with extrusion/intrusion mechanisms of fatigue initiation (Forsyth 1957) and Smith with a major model for fatigue-crack growth under high-strain conditions (Laird and Smith 1962). One aim of the present paper is to ensure that the implications of their results are fully appreciated with respect to models of LEFM crack growth.

FATIGUE LIFE DETERMINED BY CRACK PROPAGATION

For quasi-elastic (linear elastic fracture mechanics, or LEFM) conditions, the most popular, empirical relationship between the crack-growth-increment per cycle (da/dN) and parameters of stress-range ($\Delta\sigma$) and instantaneous crack length (a) is that proposed initially by Paris and Erdogan (1963) which gives:

$$da/dN = A \Delta K^m \tag{1}$$

where A and m are constants, and ΔK is the range of stress-intensity factor:

$$\Delta K = K_{max} - K_{min} \tag{2}$$

Here, K_{max} and K_{min} are respectively the maximum and minimum values of stress intensity factor in the cycle, derived from the maximum or minimum stress and an appropriate compliance function, $Y(a/W)$, which depends on loading geometry, crack length and testpiece width, W (see, for example, Knott 1973). The engineering application of equation 1) may be illustrated by considering the case of a large (infinite) body, containing a central, through-thickness crack of length 2a lying normal to a constant stress-range, $\Delta\sigma$. The expression for ΔK then becomes:

$$\Delta K = \Delta\sigma(\pi a)^{\frac{1}{2}} \tag{3}$$

so that equation 1) may be re-written as:

$$da/dN = A \Delta\sigma^m \pi^{m/2} a^{m/2} \tag{4}$$

Variables (a and N for constant $\Delta\sigma$) may be separated to give:

$$a^{-m/2} da = A \Delta\sigma^m \pi^{m/2} dN \tag{5}$$

It is then possible to integrate the two sides of this equation: the left-hand-side (LHS) between limits a_o (the initial defect size, determined by process-control or the NDT inspection limit) and a_f (the final defect size, determined by the onset of fast fracture or plastic collapse): the right-hand-side (RHS) between limits zero and the number of cycles to failure, N_f. The result of this integration, for all values of m except m = 2, is:

$$\frac{2}{(2-m)}\left| a_f^{(1-m/2)} - a_o^{(1-m/2)} \right| = A \Delta\sigma^m \pi^{m/2} N_f \tag{6}$$

For m = 2, the result is:

$$\ln(a_f/a_o) = A \Delta\sigma^2 \pi N_f \tag{7}$$

Equivalent expressions may be obtained using the compliance functions, $Y(a/W)$.

Both equations 6) and 7) express the propagation lifetime, N_f, for a given applied stress range, $\Delta\sigma$, in terms of initial and final defect sizes. Increases in N_f are expected if a_f can be increased (e.g. by using material of higher fracture toughness) or if a_o is reduced (e.g. by a process which reduces the size of initial defects). For higher values of m, N_f is particularly sensitive to the value of a_o, which may be taken as constant for a given material. Often, a_f may also be taken as approximately constant, because the number of cycles associated with growth from, say, 0.8 a_f to a_f is small compared with N_f, and the calculation of lifetime is therefore insensitive to the precise value taken for a_f. Under these circumstances, the LHS of equation 6) or 7) becomes a constant. Dividing by A and $\pi^{m/2}$, taking logarithms and rearranging, we obtain the equation:

$$\log N_f = -m \log \Delta\sigma + \text{const.} \tag{8}$$

which is essentially a linear "S-N" curve, derived from crack-growth-rate calculations. Modification to incorporate a fatigue-limit will be considered later.

The concepts underlying the derivation of equation 8) may hold more generally, even in circumstances where the details of the calculation cannot be justified because crack growth does not take place under LEFM conditions. The position of an "S-N" curve for poor-quality castings may, for example, lie to the left of that for better-quality castings, because the initial defect sizes in the poor-quality material are larger, and the lifetime for a given value of $\Delta\sigma$ is therefore smaller. Similar reasons explain improvements in the fatigue performance of offshore nodes, when welded tube-on-tube configurations are replaced by high-integrity castings, but, in this case, differences in the stress-concentration factors associated with the different types of node are also significant.

The integration procedure does not have to rely on the LEFM relationship between da/dN, $\Delta\sigma$ and crack length, as given by equation 1). It is equally possible to treat a relationship, equivalent to equation 4), of the form:

$$da/dN = A' \Delta\sigma^v a^w \tag{9}$$

where $v \neq 2w$; to separate variables, and to integrate the LHS, $a^{-w}da$, between limits a_o and a_f, to give a quantity proportional to $\Delta\sigma^v N_f$. The engineering use of an expression such as 9) is therefore precisely equivalent to that of the LEFM formulation equation 1), and the only fundamental reason for preferring the latter is if it possesses a better physical basis. Values of $Y(a/W)$ have, of course, been evaluated for a wide range of geometries, so that A in equation 1) can remain constant (although the relationship between ΔK and $\Delta\sigma$ varies) whereas the values of A' in equation 9) would have to be established for each geometry. Some of Frost's early work on fatigue-crack growth was indeed related to an equation of the form of 9) with $v = 3$, $w = 1$ (Frost, Holden and Phillips 1961).

A simple treatment (Knott 1982) of variable amplitude loading envisages a series of blocks, each of amplitude $\Delta\sigma_i$ (where $i = 1,2,3$ etc.), applied for a number of cycles, N_i, and assumes that a_f is constant for all blocks (strictly, this assumption holds only for blocks in which K_{max} is constant and variable ΔK is achieved by varying K_{min}).

Consider the first block, $\Delta\sigma_1$, during which the crack grows from length a_0 to length a_1 in N_1 cycles. The LHS of equation 5) would then be integrated between a_0 and a_1 to give an integral denoted by I_{01}. To produce failure at $\Delta\sigma_1$, the LHS is integrated between limits a_0 and a_f as in equations 6) or 7) to give I_{of}. If the lifetime at stress amplitude $\Delta\sigma_1$ is denoted by $N_{f(1)}$ it is then possible to write:

$$N_1/N_{f(1)} = I_{01}/I_{of} \tag{10}$$

Now apply a stress amplitude $\Delta\sigma_2$, for a number of cycles, N_2, during which period the crack grows from a_1 to a_2. The LHS of equation 5) now integrates to I_{12}, so that:

$$N_2/N_{f(2)} = I_{12}/I_{of} \tag{11}$$

In general
$$\Sigma \frac{N_i}{N_{f(i)}} = \frac{1}{I_{of}} (I_{01} + I_{12} + I_{23} + \ldots I_{f-1,f}) \tag{12}$$

The terms in parentheses on the RHS of equation 12) are simply the sequential integrals of $a^{-m/2}$ (or a^{-w}) from a_0 to a_1 to a_2 to a_3 . . . etc. . to a_f. By the definition of integrals their sum is equal to I_{of}, so that the resultant RHS is unity. Hence:

$$\Sigma \frac{N_i}{N_{f(i)}} = 1 \tag{13}$$

which is the Palmgren-Miner law. Although this has been derived by integrating the crack length term between sequential limits, it does not demand the LEFM formulation as such, and an expression such as that in equation 9) is equally admissible. In a more realistic treatment, the variation of a_f with $\Delta\sigma$ must be included and it is not surprising that experimental values of the Palmgren-Miner summation often differ from unity.

In summary, there are powerful engineering applications for equations which relate crack growth rate to $\Delta\sigma$ and instantaneous crack length. Initially, these may be derived in a purely empirical manner and the only justification for an LEFM formulation as such (e.g. equation 1)) is if, either, in engineering terms, it may be shown to hold for a variety of geometrical configurations better than consistent values of v and w as in equation 9), or, in scientific terms, it has a better physical basis. The first point does not seem to have been tested exhaustively: the physical basis of the LEFM relationship will be treated in later sections of the paper.

It should be noted that the form of equation 1) holds over only a limited range of crack growth-rates. Taking logarithms of both sides, it would be predicted that the graph of $\log(da/dN)$ versus $\log(\Delta K)$ should be linear, with slope m and intercept $\log A$ on the $\log (\Delta K)$ axis. Experimentally, the graph shows three regimes of which only the central regime (B) is linear (see Fig. 1). Experimental values of (da/dN) in this regime are only weakly dependent on variations in microstructure, mean stress and dilute environment.

At high values of ΔK (or, more strictly, at high values of K_{max}), the crack growth rates are higher than a simple extrapolation of the linear curve (Regime B) would predict, and we refer to regime C crack growth. The reason for this behaviour is that monotonic, or "static", modes of fracture are produced by the applied loading cycle. Usually, the importance of these monotonic modes with respect to the overall value of (monotonic + cyclic) crack growth rate depends on the values of K_{max} in the cycle, so that Regime C behaviour is a sensitive function of the mean stress, as characterised by the stress ratio, R ($R = K_{min}/K_{max}$). Since the effect results from monotonic fracture processes, the material's microstructure and fracture toughness are important. Marked changes in behaviour may be produced by altering a steel's grain size, embrittling the grain boundaries, or varying the inclusion content (Ritchie and Knott, 1973a b, Pickard, Ritchie & Knott 1975). In some early work on fatigue-crack growth, where the distinction between regimes B and C was not fully appreciated (or detected experimentally), the onset of brittle monotonic modes gave rise to apparent increases in the exponent m in equation 1) and to sensitivity of m to mean stress (see Fig. 2).

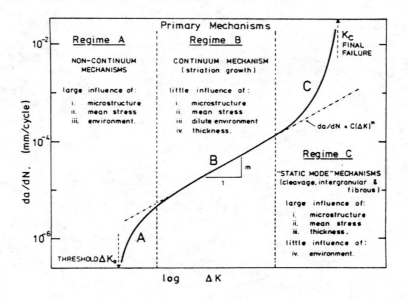

Fig. 1: Schematic Diagram indicating the Three Regimes of Quasi-Elastic (LEFM) Fatigue-Crack Growth (courtesy Prof. R O Ritchie).

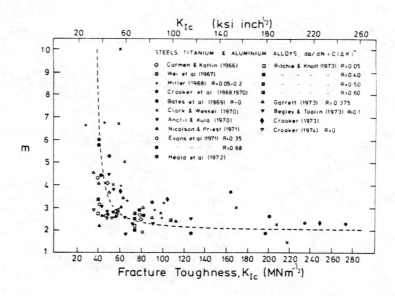

Fig. 2: Variation of the Apparent Value of the Exponent, m, in the Paris-Erdogan Equation with Monotonic Fracture Toughness (after Ritchie and Knott 1973). The references in the figure pertain to the original paper.

At low values of ΔK, the graph of log(da/dN) vs. log ΔK exhibits a cut-off, such that the value of (da/dN) decreases to zero (experimentally, to approx. 10^{-8} mm/cycle) below a "threshold" value, ΔK_{th}. Behaviour in this "near-threshold" region (Regime A) is extremely sensitive to variations in microstructure, mean stress and environment. There also are quite strong differences between the behaviour of long cracks and short cracks in the near-threshold region. The incorporation of a threshold value, ΔK_{th}, in integration procedures gives rise to a "fatigue limit" in the resulting "S-N" curve (Taylor and Knott 1982).

In the following sections, an attempt is made to provide consistent modelling of Regime B fatigue-crack-growth following the LEFM formulation, and then to indicate ways in which contributions from monotonic modes (Regime C behaviour) could be incorporated. The near-threshold region is not treated in detail, because it forms the subject of another paper in this Proceedings (Beevers 1985), but attention is drawn to experimental results which need to be included in models of threshold behaviour.

BEHAVIOUR IN SMOOTH-SPECIMEN TESTS

Thirty years ago, virtually all the fatigue tests being carried out in testing laboratories made use of smooth specimens, often subjected to rotating-bending or plane-bending, which induces stress- and strain-gradients, such that surface layers are able to deform plastically, although the specimen as a whole has not undergone general yielding. The intention of this smooth specimen testing was to attempt to characterise a material's "inherent resistance" to fatigue failure, using information from the "S-N" curve. There is, indeed, a relatively large category of engineering applications in which rather smooth bars are subjected to fatigue loading (axles, tie-rods etc.) and, here, the S-N curve may also be used to provide relevant engineering design data. It should be noted, however, that in many practical situations involving smooth bars case-hardening is employed and this must be reproduced faithfully in the testpiece, if the design data really are to be of direct application.

The results of smooth-specimen tests, however, could not be reconciled with those obtained in Structures Laboratories, where fabricated components (e.g. the "Comet" wing) were subjected to simulated service loading, using a series of hydraulic jacks or long-stroke cam/eccentric combinations. The differences lay in the fact that the smooth specimens required a number of cycles before significant (macroscopic) cracks were initiated, whereas such cracks were already present in the structural components.

It is of value to review some of the information obtained from (waisted) smooth specimen tests, because it is arguable that the material ahead of a slightly blunted, growing, fatigue crack tip is in a state similar to that on the surface of a waisted, smooth specimen, i.e. the crack tip region may simply represent a specimen of extremely short gauge-length. This gauge length could, of course, be so short that it would be smaller than the spacing of operative dislocation sources. For a value, ΔK = 9 MPa m$^{\frac{1}{2}}$ in material of flow stress 400 MPa, the "gauge length" is of order 0.25-0.5 μm, compared with an average dislocation spacing in annealed material of approximately 1.5 μm in diameter, see fig. 3 after Pickard (1977), although the cell size in fatigued material may be only 0.5 μm. This has relevance to the "damage accumulation" concepts described below.

Consider behaviour in a smooth specimen, waisted to leave a parallel-sided gauge-length of a few millimetres. For simplicity, the material is supposed initially to be annealed, single-phase, to undergo cyclic hardening and not to strain-age (i.e. not low-strength steels, or Al-Mg alloys). If sufficient stress is applied, during the first (tensile) quarter-cycle, local plasticity can be produced, even though the macroscopic stress is below the "yield stress". In the type of material specified, the conventional flow stress is the 0.2% proof stress, which requires 0.2% plastic strain. It is then quite reasonable to contemplate a smaller amount of strain at a lower applied stress and it is, in fact, extremely difficult to measure the true elastic limit, with complete absence of pre-yield microstrain in conventional tensile tests. For "push-pull", uniform loading, localised yielding is possible because the material contains extraneous stress-concentrators, such as turning-marks, scratches or fillets, and inherent stress-concentrators, such as non-metallic inclusions, carbides, dispersoids or even grain-boundary triple-points. In bending, these effects are compounded by the fact that the engineer's yield stress refers to general plastic collapse, whereas fatigue slip activity can proceed, provided simply that the surface layers are deforming plastically.

In a sense, the "ideal" model of the matrix (although not the inclusion-containing, polycrystalline material) is a pure single crystal. Results obtained on uniaxial, push-pull testpieces of copper, and copper alloys, in single crystal form, indicate

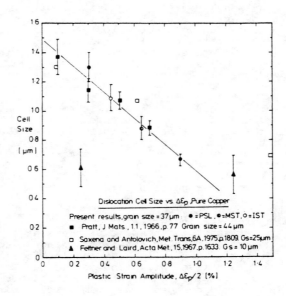

Fig. 3: Variation of Cell Size in Fatigued Copper with Plastic Strain
Amplitude (courtesy Dr. A C Pickard)

that there is a lower stress level (an inherent "fatigue limit") below which <u>co-</u>
<u>ordinated</u> dislocation movement (persistent slip bands, PSBs, composed of "walls" and
"ladders"), which eventually leads to crack initiation, does not occur (Brown 1977,
Finney and Laird 1975). The value of this stress range is quoted as approximately
± 30 MPa (R = -1) but effects of mean stress or stress-state on this value do not
appear to have been studied in detail. A critical deviatoric stress range, or
critical plastic strain range, seems more likely than a tensile stress range.
Translation from single crystal to polycrystal is not easy, because slip bands in
the real material start from stress-concentrators and hence spread under the influence of steep stress-gradients, whereas the single crystals are subjected to a uniform stress distribution. Note, however, that, even in push-pull, a large proportion of the PSBs observed is confined to surface layers (<0.2 mm deep).

At stress levels above the endurance limit (failure in 10^8 cycles), it is reasonable,
for fairly coarse-grained polycrystals, to envisage some (different amounts of)
local plasticity in each grain during the first tensile quarter-cycle, i.e. the true
elastic limit has been exceeded, although the 0.2% P.S. has not necessarily been
attained. Consider now the reverse cycle. As indicated earlier in Cottrell's (1977)
comments, slip should be geometrically reversible, so that for very low stress
amplitudes, the initial slip-bands would be linear arrays. Reversal of stress should
then simply produce a to-and-fro motion of dislocations along consecutively identical
paths, producing a slip-step offset which increased and decreased as the load changed.
Observations made by Thompson and Wadsworth (1958), Forsyth (1957) and others show
that such events are insufficient to explain fatigue-crack initiation and that a
degree of geometrical irreversibility of slip is required.

The main observations are two-fold, following from the work of Forsyth (1957).
First, it is observed that annealed material cyclically hardens and that the hardness
saturates (at least, in surface layers) before failure occurs. Secondly, some slip-
bands become increasingly intense as fatigue cycling occurs, <u>intrusions</u> occur as
shear cracks, following the slip-bands into the depth of the specimen, and these
intrusions are associated with <u>extrusions</u> at the material surface. Thin-foil trans-
mission electron microscopy indicates that the hardening results from an increase in
dislocation density as dislocations tangle with each other, forming a cell structure
if the stacking-fault energy is high, or planar arrays if the stacking-fault energy
is low. At extremely low stress amplitudes the dislocation density does not alter
significantly as cycling continues and this is attributable to a geometrical

(although not thermodynamic) reversibility. It has been proposed that dipole loops (long, parallel dislocations of opposite sign) undergo a "flip-flop" deformation, changing from one position of stable equilibrium to another as the sign of the stress changes (Feltner 1963).

Irreversibility and tangling of the sort described above requires stress levels somewhat higher than those just at the elastic limit. It is conceivable that slip-bands might have to penetrate substantially into a grain before back stresses from the grain boundary force mobile dislocations to cross-slip and tangle with network dislocations. It is often the case that the endurance limit (EL) in non-ferrous materials equals, or even exceeds, the 0.2% P.S. From the engineering point of view, the EL is still very much less than the UTS (as used in old design codes) or, in the case of bending configurations, less than the general yield stress (used in more recent codes).

Hardening in the early stages of fatigue is therefore explicable in terms of the development of tangled dislocation structures. It also seems reasonable that, as cycling continues, the dislocation structures "shake-down" to form a more-or-less stable, background configuration of cell-walls or planar arrays, with the cyclic plastic strain amplitude being carried by the movement of a small fraction of the dislocation density. This state corresponds to saturation. In plastic-strain-controlled tests, it has been shown that the cell diameter or the spacing between planar arrays is inversely proportional to the plastic strain amplitude (Pickard 1977): see Fig. 3.

The possession of a high (or low) rate of cyclic hardening cannot be equated to "good resistance to fatigue", independent of the control mode being exercised in the test. Under load-control, a high rate of hardening sequentially reduces the plastic strain amplitude per cycle, requiring less and less to-and-fro dislocation movement, perhaps reducing this to a level at which coordinated movements are not required, so that intrusions do not develop. Effects of this sort have been observed in the near-threshold region for underaged aluminium alloys (Garrett and Knott 1974). Under plastic-strain control, however, a high rate of cyclic hardening may be deleterious, because the number of dislocations moving in each cycle is dictated by the imposed plastic strain amplitude, and the effect of hardening is simply to increase tensile stress levels, which may induce brittle particles to crack, hence reducing initiation periods or accelerating crack growth. Effects of this sort may help to explain the variation of fatigue limit with mean (or maximum) stress in load-controlled tests, as demonstrated in Goodman/Soderberg/Gerber diagrams. A third control mode is that of total strain, rather than plastic strain. Here, the effect of cyclic hardening is more difficult to assess, because a low rate implies lower stresses but a larger plastic strain component, whereas a high rate is associated with higher tensile stress levels but a sequentially decreasing elastic strain component.

It is arguable that, in a non-interactive environment, a specimen under load control could cyclically harden to a saturation level and then achieve a state of geometrical reversibility, such that cracks, even if "initiated", would not grow beyond a small size. Under plastic strain control, dislocation movement is demanded each cycle, regardless of stress level, so that it is much more likely that cracks will eventually be initiated and grow. In air, adsorption of gas onto freshly exposed slip steps, or absorption of hydrogen from water vapour, precludes the condition of geometrical reversibility, so that cracks are initiated, initially from intrusions forming on 45° slip-planes, even under load control. It should be appreciated, however, that the number of cycles required to initiate a slip-band crack and propagate it across a grain diameter in a specimen under load control may require 80-90% of the total fatigue life, at least, for stresses just above the endurance limit (lifetimes of 10^7-10^8 cycles). Virtually all plastic-strain-control tests are, however, carried out for plastic strains >± 0.1% (if not >±0.2%) per cycle and crack initiation occurs in comparatively few cycles. Typical total lifetimes are approx. 10^4 cycles at ± 0.1% plastic strain.

It is often assumed that "initiation" in plastic-strain-controlled tests occurs at approximately half the lifetime. At this stage, the cyclic stress:strain curve starts to show a decrease, but, even so, initiation could have occurred earlier, if the effect of reduction of cross-sectional area on stress level (in a strain-controlled test) were masked by continued hardening of the net cross-section. For higher plastic strain amplitudes (>± 0.5%) and lifetimes < 10^3 cycles, it is arguable that the number of cycles required to initiate cracks is negligible and that the crack propagates virtually from the first cycle. It must be appreciated that these tests are usually carried out in "push-pull", so that a plastic strain of ± 0.5% implies quite large compressions. Surface grains buckle and fold in the compression cycle, producing crack-like indents by mechanisms which mirror, on the microscopic scale, the formation of geological faults by incompatible movements of tectonic plates (Laird 1962).

The preceding discussion is of value when trying to assess the relative importance of "damage accumulation" (hardening, flow localisation, intrusion formation) and "crack growth" in any fatigue failure process. A crucial point is the interpretation of the Coffin-Manson relationship (Coffin 1959, Manson 1954):

$$N_f^{\alpha} \, \Delta \epsilon_p \; = \; C \qquad\qquad\qquad (14)$$

where $\Delta \epsilon_p$ is the plastic strain amplitude, N_f is the number of cycles to failure, α is a constant (0.5-0.6), and C is a constant, obtained by extrapolation to $N_f = 0.25$ (a tensile test) and sometimes equated to the tensile fracture strain. One interpretation of equation 14) is that "damage" accumulates, cycle-by-cycle, until a critical value is achieved and failure occurs. A modification of this might be that "damage" accumulates until a number of cycles approximately $N_f/2$ is attained, at which stage a crack initiates and then propagates to failure. Supporting evidence of a sort for this view could be sought from observations that a combination of electropolishing away persistent slip-bands and annealing the specimen would restore original properties. It should, however, be emphasised that the fracture mode of transgranular fatigue propagation has nothing in common with the processes of ductile fracture, so that any connection between the constant C and the tensile fracture strain is, in general, fortuitous. Only the very final stages of plastic collapse and fracture of the uncracked ligament during the final tensile quarter-cycle occur by processes similar to those in the tensile test. Any "damage accumulation" argument must therefore relate to the development of propagating cracks, which, in turn, progressively reduce the cross-sectional area until failure occurs. The nature of the damage is two-fold: a) the saturation of hardening, to allow flow to remain localised, b) the development of persistent slip-bands to produce propagating cracks.

The alternative interpretation of equation 14) is that it reflects the progressive growth of a single crack across the cross-section. This, of course, assumes that the "initiation" stage (hardening, PSB formation) is negligible, which is particularly true for the higher strain amplitudes in fully-reversed loading. A model developed by Tomkins (1968) relates the crack growth increment per cycle, da/dN, to the crack-opening-displacement range at the instantaneous crack tip, $\Delta \delta$, generated by the applied plastic strain amplitude, $\Delta \epsilon_p$. The cross-section is assumed to be saturation-hardened to a uniform flow stress, σ_f, and the plastic deformation associated with the new crack-opening is represented as a single slip-line traversing the section. The relationship between da/dN and $\Delta \epsilon_p$ is derived as:

$$\frac{da}{dN} \; = \; \frac{\pi}{8(2\sigma_f)^2} \, \Delta \epsilon_p (\Delta K)^2 \qquad\qquad (15)$$

and this may be integrated in a manner similar to that employed for equations 4) and 9), assuming the cyclic stress/strain law:

$$\Delta \sigma \; = \; k \, \Delta \epsilon_p^{\beta} \qquad\qquad\qquad (16)$$

to give the expression:

$$\Delta \epsilon_p \, N_f^{\{1/(2\beta+1)\}} \; = \; \left| \frac{8}{\pi^2} \left(\frac{2\sigma_f}{k} \right)^2 \, \ln \left(\frac{a_f}{a_o} \right) \right|$$

$$= \; \text{const.} \qquad\qquad\qquad (17)$$

cf. equation 7). This is clearly similar in form to the Coffin-Manson equation 14).

The physical mechanisms involved in Tomkins' model are closely similar to those involved by Laird and Smith (1962) following their studies of fully-plastic crack growth. The importance of crack-opening-displacement range $\Delta \delta$ may be appreciated by considering an initially oxidised, or filmed, crack. During the tensile quarter-cycle, the new plastic deformation emanates from the crack tip, producing clean, unfilmed, slip steps which accommodate the crack opening. Gaseous species from the environment can then adsorb onto the clean, reactive surfaces and prevent slip reversibility during the rest of the cycle. If the supply of gaseous species is plentiful, the number of atoms adsorbed in a given time will depend on the amount of fresh area exposed, i.e. for unit thickness, on the magnitude of $\Delta \delta$. As more atoms are adsorbed, at high $\Delta \delta$, the absolute magnitude of the non-reversible component increases, so that the value of da/dN increases with $\Delta \delta$. Similar arguments apply if chemical interaction with the gaseous species occurs (e.g. hydrogen uptake and embrittlement), provided that the absolute amount of interaction is again dependent on the amount of fresh area exposed in the slip-band.

This physical picture of crack growth dependent on the amount of non-reversibility of slip produced by interaction with the environment plausibly applies also to LEFM growth, where it is possible to obtain a simple relationship between $\Delta\delta$ and K. This will be the subject of the following section. Here, it is of more significance to examine the implications of Tomkins's analysis with respect to the Coffin-Manson equation and concepts of damage accumulation. The good agreement found between equations 17) and 14) strongly suggests that the major component of "damage" is crack-growth (under plastic strain control) rather than "pre-initiation" hardening or PSB development, although it is arguable that the growth of PSBs is also a simple function of plastic strain amplitude, so that PSB growth is virtually indistinguishable from the growth of a small crack. At high strain amplitudes, the folding and buckling in compression effectively eliminates any initiation period. At high $\Delta\epsilon_p$, therefore, the physical plausibility and analytical agreement given by Tomkins's model tends to eliminate* significant effects of damage accumulation other than crack growth.

LEFM CRACK GROWTH

The main differences between crack growth in the Coffin-Manson test and that under LEFM conditions are that: firstly, the Coffin-Manson testpiece is subjected to macroscopic plastic displacements, whereas the LEFM testpiece has its plasticity limited to small zones surrounded by an elastic matrix; secondly, the Coffin-Manson testpiece is subjected to through-zero cycling, so that plastic displacements are produced in compression as well as in tension, whereas the LEFM testpiece is cycled at zero or positive stress-ratio. Compressive displacements tend to facilitate initiation as described above, but, presumably, do not contribute to long crack growth, because increasing compression simply presses the crack flanks together, more tightly. In LEFM growth, it seems appropriate to concentrate simply on the tensile crack-opening-displacement range, $\Delta\delta$. Initially, this can be calculated from applied values of stress-intensity factor, but under some conditions, particularly those of decreasing stress-intensity and low stress-ratio ($R = K_{min}/K_{max}$), it is necessary also to consider aspects of crack closure, which can reduce the effective crack-tip opening displacement range.

Under monotonic tensile loading in LEFM or "small-scale yielding" conditions, the values of crack-opening-displacement, δ_{max} or δ_{min} corresponding to stress-intensity factors, K_{max} or K_{min} are given by (see e.g. Knott 1973):

$$\delta_{max} \approx K_{max}^2/2\sigma_Y E; \quad \delta_{min} \approx K_{min}^2/2\sigma_Y E \qquad (18)$$

where σ_Y is the yield stress and E is Young's modulus. Under cyclic loading, the size of the reversed plastic zone, and hence $\Delta\delta$, in non-hardening material is obtained by substituting $2\sigma_Y$ for σ_Y, and if cyclic hardening occurs in the plastic zone, σ_Y should be replaced by the cyclic flow stress, σ_f. Hence:

$$\Delta\delta = \delta_{max} - \delta_{min} \approx (K_{max}^2 - K_{min}^2)/4\sigma_f E \qquad (19)$$

Writing $R = K_{min}/K_{max}$, $\Delta K = K_{max}-K_{min}$, $K_{max} = \Delta K/(1-R)$, $K_{min} = R\Delta K/(1-R)$, we have

$$\Delta\delta = \{(1+R)/(1-R)\} \Delta K^2/4\sigma_f E \qquad (20)$$

If it is now assumed that the crack-growth increment per cycle, da/dN, is linearly proportional to $\Delta\delta$, (following a gas-adsorption argument, or by analogy with Tomkins's model for fully plastic crack growth equation 15) equation 20) predicts that growth in Regime B should depend on stress-ratio through the form $(1+R)/(1-R)$ and that the exponent, m, in the Paris-Erdogan law (equation 1)) should be equal to 2. There is also a suggestion that growth-rates in different materials should scale inversely with Young's modulus, but the details of interaction between environment and clean metal surfaces in materials with markedly different moduli (steels, aluminium alloys, titanium alloys) could be so different that they would mask the simple mechanical effect of modulus. Experiments in hard vacua are needed.

Rather few studies of effects of stress ratio on da/dN have been made in materials for which it is clear that only Regime B growth is occurring, but fig. 4 shows results obtained by El Soudani (1981) and by Pickard (1975) for 316 austenitic stainless steel at appropriate ΔK levels. A good linear relationship between da/dN and the term $(1+R)/(1-R)$ is obtained for a constant value of $\Delta K = 22.5$ MPa m$^{\frac{1}{2}}$, and it would be possible to deduce that a linear relationship held also at $\Delta K = 36$ MPa m$^{\frac{1}{2}}$, although here, there is rather more scatter observed in the individual values of da/dN. It should be re-emphasised that these results, particularly for $\Delta K = 22.5$ MPa m$^{\frac{1}{2}}$, refer to crack growth dominated by the Regime B, striation mode. In commercial aluminium alloys, for example, growth at similar ΔK levels would be affected strongly by the occurrence of "static" modes associated with intermetallic,

---

*provides "the last nail in the coffin" for

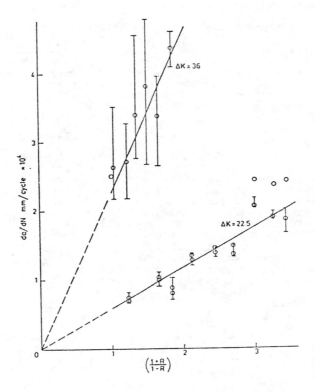

Fig. 4: Variation of da/dN in 316 Stainless Steel with the Parameter
{(1+R)/(1-R)}, following Equation 20). (Data courtesy
Dr. S M El Soudani and Dr. A C Pickard).

"dispersoid" particles.

The importance of "static" modes of fracture, with respect to conventional, macroscopic measurements of da/dN, was high-lighted initially by the work of Ritchie and
Knott (1973). Arguing that the techniques used to measure crack growth (potential
drop, optical, compliance) gave values representing total (cyclic plus static)
growth, they were able to show major differences in growth-rate induced by treatments such as temper-embrittlement of alloy steel or coarsening the grain-size in
mild steel. In both cases, "bursts" of brittle (intergranular or transgranular)
fracture produced discrete increments in total crack growth rate, which could be
detected by the sensitive potential drop system employed, but which would have been
perceived by less precise systems simply as an overall increase in da/dN, implying
higher apparent values of the exponent m in equation 1). An examination of results
quoted in the literature, reproduced in fig. 2, showed that it was possible to
relate values of m to a material's __monotonic__ fracture toughness, K_{IC}. For tough
materials, where brittle, "static" modes are unlikely, the values of m are generally
close to 2, in accordance with the simple model based on $\Delta\delta$. Values of m substantially higher than 2 are observed only for materials of low toughness (low K_{IC}),
where it would be expected that increments of brittle crack growth could occur
during the fatigue test, particularly at high stress-ratio.

In addition to the value of these observations with respect to interpretation of
behaviour in fatigue-crack growth-rate testing, there are practical implications,
particularly related to the behaviour of welded joints. In non-stress-relieved
welds made in high-strength, low-alloy (C/Mn) structural steels, it is possible to
obtain coarse grains in the heat-affected-zone (HAZ) and these can have low toughness at sub-ambient temperatures (perhaps as low as -40°C in arctic applications).
Additionally, thermal contraction of the weld metal gives rise to high tensile
residual stresses, which implies that any fatigue cycles are associated with high
R-ratios. Cleavage bursts have, indeed, been observed as a part of "fatigue" crack
growth in such steels.

For stress-relieved alloy steel weldments, on the other hand, slow-cooling from the
stress-relief temperature could induce intergranular segregation of trace impurity
elements (P,Sb,Sn), which would tend to render the steel more susceptible to intergranular fracture. The level of residual stress would, of course, now be much
reduced, so that high stress ratios could arise only from the service loading

spectrum. Recent evidence has been obtained for dynamic segregation during the stress-relief heating cycle and this can produce strong effects of dwell periods and frequencies during fatigue crack growth at temperatures of approximately $500^\circ C$ (Bowen 1985).

A simple model of effects of static modes on crack growth-rate makes use of results obtained by Beevers et al. 1975 which relate the area percentage of cleavage facets observed on the fracture surface to the associated value of K_{max} during the cycle (figs. 5,6). As K_{max} approaches K_{IC}, the area <u>fraction</u>, A_C approaches unity in a more-or-less linear fashion. Assuming that this is linear, we may write:

$$(1 - A_C) = k(K_{IC} - K_{max}) \tag{21}$$

where k is a constant of proportionality. Now consider a band of material, of area, B (thickness) x δa(crack extension) subject to given values of ΔK and K_{max}. If there were no cleavage, the number of cycles N_O required to traverse the area $B\delta a$ would be $N_O = \delta a/(da/dN)$, since any increment of crack growth is equated to an equivalent linear extension multiplied by testpiece thickness, B. Now assume that every cleavage component occurs virtually instantaneously (each "burst" requiring no more than a quarter of a cycle). The area which has to be traversed by fatigue is then reduced to $B\delta a(1-A_C)$ so that, for the same ΔK and hence the same true value of (da/dN), the number of cycles required completely to traverse the reduced area, N_1, is reduced to $N_1 = (1-A_C)\delta a/(da/dN)$. The apparent value of da/dN, $(da/dN)_{app} = \delta a/N_1 = (\delta a/N_O)/(1-A_C)$ is then increased with respect to the true value, (da/dN) by the factor $(1-A_C)^{-1}$. From equation 21), this may be written as $k(K_{IC}-K_{max})^{-1}$, so that the overall expression for $(da/dN)_{app}$ becomes:

$$\left(\frac{da}{dN}\right)_{app} = \frac{A''\Delta K^2}{4\sigma_f E}\left(\frac{1+R}{1-R}\right)\left(\frac{1}{K_{IC} - K_{max}}\right) \tag{22}$$

or, writing

$$K_{max} = \Delta K/(1-R)$$

$$\left(\frac{da}{dN}\right)_{app} = \frac{A''\Delta K^2 (1+R)}{4\sigma_f E\{(1-R)K_{IC} - \Delta K\}} \tag{23}$$

where A" incorporates the constant k from equation 21). The expressions in equations 22) and 23) bear fairly close resemblance to those used by McEvily (1977) and by Forman (1967), but have been derived in a manner intended to represent physical behaviour. Effects of stress-ratio should be recognized to be of two kinds: a) they affect the relationship between $\Delta\delta$ and ΔK^2 (equation 19); b) the value of K_{max} controls the onset of static brittle modes.

Clearly, the value of k in equation 21) is an empirical, experimental parameter: from figs. 5 and 6, the figures are 0.055 and 0.065 respectively. A more scientific interpretation requires a statistical analysis of the probability of a given size and orientation of cleavage microcrack nucleus propagating under the influence of a given value of K_{max}. Models based on monotonically increasing stress intensity would suggest a dependence of A_C on K_{max} much steeper than linear (e.g. on K_{max}^4, because this determines the highly-stressed volume ahead of a crack tip), but the situation in fatigue is complicated by the fact that K_{max} decreases rapidly during the unloading part of the cycle, so that the microcrack does not propagate over large distances. Behaviour will be affected by the frequency and wave-shape of the fatigue cycle (which controls the time for which high stress intensity is applied) and by the strain-rate response of the matrix.

Void growth in the plane-strain, central sections of fatigue testpieces made in 316 stainless steel weld metal (which contains a high volume fraction of closely-spaced non-metallic inclusions) also appears to be a function of K_{max}: see figs. 7,8 (Pickard et al. 1975, Knott 1981, El Soudani 1980). It is clear that the overall crack-growth rate is increased by the presence of the voids, but the simple treatment employed for cleavage bursts is unlikely to hold, unless the voids form very rapidly compared with the fatigue crack growth-rate. This is perhaps more likely in aluminium alloys, which contain rather brittle intermetallic, dispersoid particles.

For striation growth, the arguments have rested on the value of $\Delta\delta$ at the tip of the propagating crack, and it should be made clear that this type of growth (Regime B) holds over only a limited range of conditions, where neither static modes (Regime C) nor closure effects in the threshold region (see the following section) operate. An alternative interpretation of Regime B behaviour has been that of "damage accumulation" (Weertman 1969). By analogy with behaviour in smooth-specimen, plastic-strain-controlled tests, the "gauge length" ahead of the crack tip in an LEFM specimen is supposed to accumulate "damage" until it is "conditioned" to a state in which it is

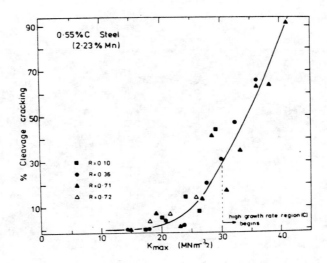

Fig. 5: Dependence of Area Percentage of Cleavage Facets with K_{max} (after Beevers et al. 1975).

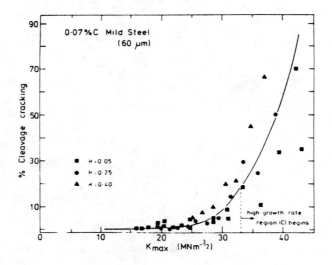

Fig. 6: As Fig. 5, but for O.07%C Steel (after Beevers et al, 1975).

44

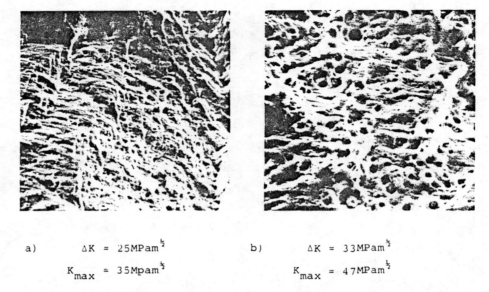

a) $\Delta K = 25 \text{MPam}^{\frac{1}{2}}$ b) $\Delta K = 33 \text{MPam}^{\frac{1}{2}}$

$K_{max} = 35 \text{Mpam}^{\frac{1}{2}}$ $K_{max} = 47 \text{MPam}^{\frac{1}{2}}$

Fig. 7:   Mixed Voids and Striations in Fatigued 316 Stainless Steel Weld
          Metal (courtesy Dr. A C Pickard).

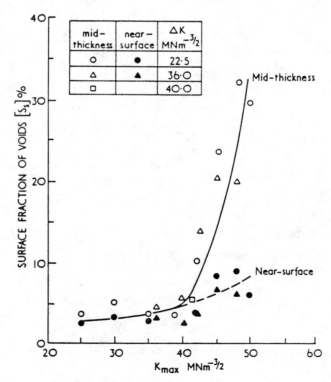

Fig. 8:   Dependence of Area Percentage of Voids on K_{max} in Fatigued 316
          Stainless Steel Weld Metal;   data courtesy Dr. S M El Soudani -
          see also Knott (1980).

prone to fracture. The crack then advances by an "increment" and the damage-accumulation process has to be repeated. The initial motivation for damage accumulation models arose from early quoted values for the exponent, m, in the Paris-Erdogan equation (1) of 4. To account for this value, it was assumed that a critical level of plastic strain had to be achieved throughout an area (volume) ahead of the crack tip which was proportional to the square of the extent of the plastic zone, i.e. to ΔK^4. As shown above, the value of m for tough materials in Regime B is, in general, equal to 2 rather than to 4 (see fig. 2), and, although this conclusion will be qualified below, the observation tends to remove the necessity for a strain-accumulation model. The main problem with damage-accumulation concepts, however, comes from the results of annealing experiments, which have been carried out both in aluminium-copper alloys and in 316 stainless steel. Results for stainless steel obtained by Pickard (1977) are shown in fig. 9, and these show clearly that the effect of a high-temperature (1100°C) anneal is to cause a transitory increase in da/dN (see also Garrett and Knott 1976). On the damage accumulation argument, the anneal would remove the damage accumulated during previous fatigue cycles, so that a delay or retardation in (da/dN) would be expected, until sufficient new damage had been accumulated to "condition" the crack-tip "gauge-length" to enable it to fracture. This runs completely contrary to the observed increase in (da/dN) which may be explained readily by the $\Delta\delta$ model for cyclically-hardening materials such as 316 stainless steel). If the fatigue has produced a region of cyclic hardening ahead of the crack tip ($\sigma_f > \sigma_Y$), the value of $\Delta\delta$ in steady-state growth is inversely proportional to the cyclic flow stress, σ_f (equation 19)). After annealing, the cyclically-hardened region is effectively removed and the value of σ_f drops to that of the yield stress, σ_Y, until cyclically-hardened structures re-develop. Consequently, the transitory value of $\Delta\delta$ increases, because $\sigma_Y < \sigma_f$, but returns to its original value (at constant ΔK) once the cyclic zone is fully re-established. Since it is assumed that (da/dN) is proportional to $\Delta\delta$, it would be predicted that (da/dN) would initially increase and then return to its original value, as observed.

The conditions leading to m = 2 in engineering materials are not always easy to meet: in aluminium alloys, for example, there is very little growth that is not either in the threshold region or in Regime C. Another point involves the techniques used to measure crack length, and the principle may be best illustrated by considering polycrystalline material in which fatigue-crack growth is affected by the crystallographic orientations of individual grains. Suppose that, in a testpiece of

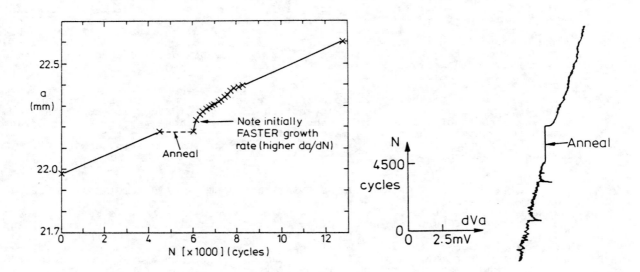

Fig. 9: a) Effect of Annealing on Fatigue Crack Growth in 316 Stainless
         Steel (Regime B). b) P.D. Trace (courtesy Dr. A C Pickard)

thickness B = 25 mm, and grain diameter 25 μm, there are 1000 grains along the line of the crack front. At a low ΔK, suppose that 100 of these grains are oriented so as to be able to sustain a growth rate of 1 μm per cycle, and that the growth-rate in the other 900 grains is negligible in comparison. The potential drop technique measures the change in electrical resistance and hence the change in crack area. This area is then divided by B to obtain the equivalent crack extension; with only 10% of the crack front growing, the apparent growth-rate would be 0.1 μm per cycle. Now suppose that ΔK is increased. The 100 grains will exhibit faster growth (da/dN ∝ ΔK^2), but, in addition, more grains will be activated, because the crack-tip stresses are generally higher, so that the measured growth rate shows a steeper dependence on ΔK. For example, the 100 grains may grow at 2 μm per cycle, and a further 200 grains may grow at 1 μm per cycle, with the other 700 grains effectively dormant. As a result of the measurement and averaging procedures, the apparent rate at the higher ΔK value would not be 0.2 μm per cycle (100 grains at 2 μm per cycle) but 0.4 μm per cycle (100 grains at 2 μm per cycle plus 200 grains at 1 μm per cycle). In practice, of course, the growth will not divide up in such a discrete manner, but the general principle holds. The effect is likely to be particularly marked at low ΔK, and it is, indeed in such regions that microscopic growth rates, measured from fatigue surfaces, are found to be substantially greater than macroscopic rates, but to correspond to exponents close to 2, compared with higher exponents for the macroscopic rates. Figures for 316 stainless steel (Pickard, Ritchie and Knott 1975) are:

$$\Delta K = 28 \text{ MPa m}^{\frac{1}{2}}$$

microscopic rate = 0.52 μm/cycle
macroscopic rate = 0.18 μm/cycle

$$\Delta K = 50 \text{ MPa m}^{\frac{1}{2}}$$

microscopic rate = 1.20 μm/cycle
macroscopic rate = 1.00 μm/cycle

Growth at higher ΔK, particularly in materials such as quenched-and-tempered steels, is likely to be far more uniform and should give values of m close to 2, as shown in fig. 2.

The final point concerning the Δδ model concerns the effect of environment, since gas adsorption on freshly-exposed slip-steps at the crack tip has been invoked to explain the lack of geometric reversibility. Growth rates in vacuum are certainly less than those in air at low ΔK, but fatigue cracks can grow even in a fairly hard vacuum (∼ 10^{-7} mbar) and, at high ΔK, little difference is observed between rates in air and those in vacuum (Kendall and Knott 1984b). Excluding strong contributions from static modes at high ΔK, it appears possible that mechanical irreversibility can be produced simply by tangling "forward-moving" dislocations into cell structures or causing them to cross-slip because back-stresses (e.g. from network dislocations or grain-boundaries) are high. If "backward-moving" dislocations then emerge on different planes, geometrical notches will be produced (as in the Cottrell-Hull (1957) intrusion mechanism) and continuing growth could arise from progressive "notch" formation. It is, of course, possible that highly-dislocated cell walls simply break apart by tensile fracture to give crack advance, but this would imply sensitivity to K$_{max}$ and lower growth rates after annealing, because the cell walls would be eliminated. Sensitivity to K$_{max}$ seems not to occur and effects of annealing on growth-rates in vacuum have not been explored in detail, although preliminary results suggest higher, rather than lower, rates.

The third region of crack growth is that near the threshold. This topic will be treated in detail in another article in the present volume (Beevers 1985) and the brief account given in the following section is simply for completeness, and to contrast effects of closure in threshold testing from those associated with crack growth in Regimes B and C. One major difference is that most fatigue crack growth occurs under conditions of increasing ΔK whereas threshold testing employs decreasing ΔK sequences.

FATIGUE CRACK GROWTH IN THE "NEAR-THRESHOLD" REGION (REGIME A)

Two main features of fatigue-crack growth at low values of ΔK will be discussed: the occurrence of "crystallographic" or "microstructurally sensitive" growth, and the main factors associated with the "threshold", ΔK$_{th}$. Crystallographic growth has been observed particularly in titanium alloys (Beevers 1977) and in f.c.c. alloys which have rather large grain size and in which the deformation is concentrated on primary slip planes, i.e. cross-slip is limited. Examples of f.c.c. alloys displaying crystallographic growth are zone-hardened, underaged Al-Cu alloys (Garrett and Knott 1975) nickel-aluminium bronze (Pickard, Ritchie and Knott 1976), Cu-Al alloys (Higo, Pickard and Knott 1981) and austenitic stainless steel (Pickard et al. 1975), all possessing low stacking-fault energy; and nickel-base superalloys which are hardened by ordered γ′ precipitates (King 1981).

In the f.c.c. alloys, there has been some question as to whether the crystallographic facets observed are simply {111} slip-planes or whether they are {001} planes, as indicated (macroscopically) for some facets, using techniques such as etch-pitting or micro-focus Laue X-ray back reflection to determine the orientation (see fig. 10). There are two points to make. First, in thin testpieces, flow is likely to concentrate on a single set of through-thickness slip-planes, because the minimum principal stress is the through-thickness stress. Observations made on thin sheet or at the edges of a thick testpiece are therefore likely to reveal facets coinciding with {111} planes, in agreement with experiments.

The plane strain opening of a crack in the central section of a thick testpiece is, however, effected in a continuum by a plastic zone having the form of two lobes symmetrically inclined at $\pm 70^{\circ}$ to the line of crack extension. If, in an f.c.c. alloy, the crack lies parallel to a macroscopic {001} plane, and is normal to the applied tensile stress, a similar opening may be achieved by the movement of dislocations, of Burgers vector $(a/2)<1\bar{1}0>$ on two appropriate {111} slip-planes, symmetrically inclined to the line of crack extension. The net effective opening may be written as if it were a dislocation reaction. Consider a crack whose plane is parallel to (001) propagating in the direction [110], so that the line of the crack front is $[1\bar{1}0]$. The two slip-planes are $(11\bar{1})$ and $(1\bar{1}\bar{1})$ which intersect along $[1\bar{1}0]$ and the slip dislocations have Burgers vectors $(a/2)[0\bar{1}1]$, glissile in $(11\bar{1})$, and $(a/2)[011]$, glissile in $(1\bar{1}\bar{1})$ respectively. The reaction is then

$$a[001] = (a/2)[0\bar{1}1] + (a/2)[011] \qquad (24)$$

showing that a displacement normal to (001) could be produced by these symmetrical slips. In practice, it is found that the macroscopically flat facets are composed of fine-scale ridged striations, parallel to the $[1\bar{1}0]$ direction of the crack front, and it appears that the opening is achieved by sequential slips on the $(11\bar{1})$ and $(1\bar{1}\bar{1})$ planes, rather than by tensile separation in the [001] direction. Certainly, the rate of crack growth is dependent on ΔK (i.e. to-and-fro dislocation movement producing the crack-tip displacement range $\Delta\delta$) rather than on K_{max}, so that the apparent "cleavage" on {001} planes in these f.c.c. alloys is of a character very different from that observed for brittle cleavage in b.c.c. mild steel, described earlier. It should be noted that the resolution of the scanning electron microscope is such that the best resolution that can be achieved is of order 0.1 μm (1 mm at x 10,000) so that the absence of obvious striations at (local) growth-rates of less than 10^{-4} mm/cycle cannot provide any support for simple tensile separation.

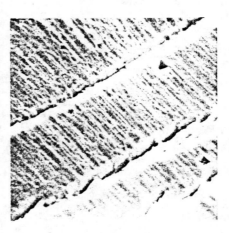

a) macroscopic {001} facets in aluminium-copper (after Garrett and Knott 1974).

b) ridged facets in copper-8% aluminium (after Higo et al.1981). Note etch pit, whose form is compatible with a [110] propagation direction and a $[1\bar{1}0]$ line of the crack front.

Fig. 10: Crystallographic Facets in Fatigue.

In nickel-base superalloys, hardened by ordered γ' precipitates, the individual slips on the two {111} planes are large, so that, at low ΔK, the fracture surface shows pronounced facetting, which is composed predominantly of large segments of {111} planes (aligned in "plane strain" rather than "plane stress" configurations), although small regions show macroscopic {001} facets (King 1981). The criterion for facetting is based on the fact that the plastic zone size must be less than the material's grain size: due to the large slips, this implies that the fatigue surface for coarse-grained material at low ΔK is much "rougher" than that for fine-grained material. Similar effects are found in titanium alloys.

It is appropriate at this point to comment on the threshold value, ΔK_{th}. This is determined conventionally by starting a fatigue crack at a relatively high level of ΔK and reducing this in 10% or 5% decrements each time the crack has grown through a distance corresponding perhaps to four times the size of the plastic zone corresponding to the previous ΔK level. When no crack growth can be detected in 10^6 cycles (with 10 μm resolution, this corresponds to a growth rate of less than 10^{-8} mm/cycle), the value of ΔK is equated to the threshold ΔK_{th}. It is usual to "grow the crack out" from the threshold, to ensure that the growth-rate curves for decreasing ΔK and increasing ΔK superimpose but, in relatively small specimens, it is not usually possible to make more than one measurement of ΔK_{th} per specimen.

The experimental values of ΔK_{th} are found to vary widely with stress-ratio, R., with environment, and with microstructure. The reasons for this are discussed in detail elsewhere (Ritchie 1981, Beevers 1985), but a brief summary of points is as follows. On the basis of a Δδ model for fatigue-crack growth (McEvily and Groeger 1977) we write:

$$da/dN \propto \Delta\delta \{(1+R)/(1-R)\} \Delta K^2 \tag{25}$$

where the constant of proportionality between da/dN and Δδ depends on the environment (availability of gaseous species for adsorption; chemical attack). If the threshold ΔK_{th} is now associated with a critical minimum value of Δδ (sufficient perhaps to generate sufficient to-and-fro dislocation movement to penetrate protective films at the crack tip), we may write

$$\Delta K_{th}^2 \propto \{(1-R)/(1+R)\} \Delta\delta_{min} \tag{26}$$

If ΔK_O is then the value of K_{th} when R = 0, we have

$$\Delta K_{th} \propto \{(1-R)/(1+R)\}^{\frac{1}{2}} \Delta K_O \tag{27}$$

which predicts the form of the stress-ratio dependence. This equation fits experimental data rather well, except at high R values, where crack-tip blunting may be of importance. In view of the closure arguments described below, the agreement at low R is surprisingly good.

The main alternative reason put forward to explain the stress-ratio dependence of ΔK_{th} is that, even in plane strain, the crack flanks in low R tests make contact prematurely at a stress intensity, K_{op}, greater than K_{min} in the applied cycle, so that ΔK is reduced to an effective value, ΔK_{eff}, given by:

$$\Delta K_{eff} = K_{max} - K_{op} \tag{28}$$

The term K_{op} is used because the "closure" load is usually measured as an "opening" load, after a cracked specimen has been fully unloaded and reloaded. Several causes for premature contact, or "closure", have been proposed: of the three treated briefly here, two (surface-roughness-induced closure and oxide-induced closure) are supposed to operate close to (< 0.1 mm behind) the crack tip; the third (plastic-wake-induced closure) may produce an apparent shortening of the crack by some 1-5 mm.

The importance of the plastic-wake and the decreasing-ΔK sequence employed in threshold testing has recently been emphasised by James and Knott (1985a). They showed that, at low R (0.2 and 0.35), there was good evidence for closure well behind (≥ 1 mm) the crack tip and that the increase in the ratio K_{op}/K_{max} observed at low ΔK could be predicted from "steady-state" results at high K_{max}, simply by calculating the "steady-state" ratio for a value of K_{max} in the wake corresponding to the "closed" 1 mm or so behind the crack tip. Spark-machining into the crack to remove plastic wake produced a decrease in K_{op}/K_{max} and a corresponding increase in da/dN, and similar effects would be expected if "short cracks" were prepared by machining away material (conditioned by the plastic wake) from behind the tip of a long crack. It is reasonable that one of the causes of closure should be the plastic wake, because this is comprised of a set of dislocation arrays which tend to exert compressive stresses behind the crack. From the practical point of view, however, doubt is cast on the value of ΔK_{th} as a design parameter for defects which are present in a structure but which are not associated with the same plastic-wake history as that employed in the test procedure. Such defects might be shrinkage cavities in castings, or hydrogen cracks in welds. For these, values of ΔK_{th} measured at low R could be unrealistically high and if the cracks were then to grow under conditions of increasing ΔK, design based on ΔK_{th} would be non-conservative. Possible procedures are to use values of ΔK_{th} obtained at high R (assuming that monotonic blunting does not occur) or to make use of a critical value of ΔK_{eff} in equation 28). In the experiments of James and Knott (1985b), results at all values of R (in the range 0.2-0.7) could be reconciled by the use of a critical value, $\Delta K_C \simeq 3$ MPa m$^{\frac{1}{2}}$ in air; $\Delta K_C \simeq 7$ MPa m$^{\frac{1}{2}}$ in vacuum.

Although the plastic wake provides general motivation for closure behind the crack tip and could give rise to closure even for crack flanks which were smooth and clean, further effects can be produced by rough surfaces or by the formation of hydrated oxide just behind (< 0.1 mm) the crack tip (Ritchie 1981). In a material such as a nickel-base superalloy, the crack opening at any point is essentially asymmetrical, so far as the plastic component is concerned, because the majority of the slip proceeds on a single {111} plane. Even in the absence of a plastic wake premature closure could then occur by asperities touching before K_{min} was obtained. As K increases, the crack opening comprises elastic opening, normal to the crack plane, plus a shear along a {111} plane. During the unloading from K_{max}, the elastic recovery simply closes the crack, but the amount of reverse plasticity is much less than that in the forward direction, so that, if the fatigue surface is steeply ridged (traces of {111} planes), contact can be made before the stress intensity is reduced to K_{min}. A similar effect of surface roughness appears to explain the dependence of ΔK_{th} on grain size in mild steel for both air and vacuum environments (Kendall and Knott 1984a).

The role of a gaseous environment is subtle. Values of ΔK_{th} are substantially higher in a hard vacuum than in laboratory air, but values in air are higher than those in a dry, inert gas, such as helium. The general effect of air vs. vacuum on growth-rate is to be expected from arguments based on the prevention of crack-tip re-welding by adsorption of gaseous species, although it is difficult to see how this can be extended to treat the threshold values as such, when tests are carried out with R-ratios as high as 0.5, and the differences between air and dry gas then appear to be anomalous. The reason is that, for steel in relatively moist laboratory air, it is possible, at low ΔK, to produce rapid growth of a hydrated oxide just behind the crack tip. This effectively wedges the crack open, so that the full value of ΔK is not experienced. This growth of oxide cannot occur in dry, inert gas, and so the value of ΔK_{th} is lower than that in air. Gas adsorption can still take place on freshly exposed surfaces, however, so that growth rates are much higher than those in vacuum.

It is clear, from the points made above, that behaviour in the threshold region is a topic of scientific interest and engineering importance. The causes of closure need to be clarified further and it is necessary, in particular, to decide whether closure is occurring just behind (< 0.1 mm) or well behind (~ 1 mm) the crack tip. This bears strongly on the sensitivity of technique used to measure closure. At present, there is concern that some of the values of ΔK_{th} measured at low stress ratios may be artificially high, compared with those that would be appropriate for natural defects in service, as a result of the load-shedding procedure followed in the threshold test. This would contribute also to discrepancies between growth-rates measured for short cracks at low ΔK and those measured for long cracks. It seems prudent, for design, to employ high R values or a critical value, $\Delta K_C \simeq 3$ MPa m$^{\frac{1}{2}}$ in air, until the relevance of closure to the growth of realistic service defects has been established. Such a study would have to involve the residual stress distribution present in service, because any compressive components act in a manner similar to that of the plastic wake in reducing the value of ΔK_{eff}.

CONCLUSIONS

The paper has tried to present a consistent, physically-based, set of models for the processes involved in fatigue-crack propagation under quasi-elastic conditions in real engineering alloys. In essence, it has been argued that the fatigue processes are primarily concerned with alternating slip at the crack tip, but that the actual crack growth increment per cycle is affected by lack of re-welding, either by gaseous adsorption or by chemical interaction with freshly exposed material at the crack tip. These processes are a natural extension of the observations made by Forsyth, Smith and others. Equating the freshly exposed area per unit thickness to the crack-tip opening displacement range, $\Delta\delta$, gives an exponent m = 2 in the Paris law, and it has been shown experimentally that this is the most appropriate value for the mid-range of growth rates. At high values of K_{max}, "static" fracture modes are involved as part of the materials response to the applied load cycle and a simple model to treat these is proposed. Factors affecting growth-rates in the near-threshold region have been briefly discussed, drawing attention to crystallographic growth and to mechanisms of closure which affect the threshold value, ΔK_{th}. For design purposes in steel structures, our present lack of understanding suggests that it would be prudent to employ a value of $\Delta K_c \approx 3$ MPa m$^{\frac{1}{2}}$ in air, although a higher value appears to be appropriate for vacuum conditions (buried cracks).

In the space available, it has not been possible to treat every aspect of fatigue-crack growth and it is, perhaps, of interest finally to draw attention briefly to some of the topics that have been omitted and to suggest areas in which useful experiments can be made. The first of these topics is the effects of programmed loading sequences or single overloads on crack growth. Often, retardation is observed, and this affects the Miner's Law summation in equation 13) in a benign manner, but it is important to understand why the retardations occur, and to be sure that deleterious accelerations cannot occur (e.g. by "static modes" in response to a high K_{max}). In a tough material, enhanced closure as a result of overloads may be produced, but effects of ΔK independent of R-ratio have also been observed (Knott and Pickard 1977) so that the relative importance of a number of factors needs to be established.

A second area concerns the general effect of environment on fatigue crack growth. In aqueous solutions, it is important to realise that the opening and closing of a fatigue crack not only exposes clean surface at the crack tip on each cycle, but also flushes out spent solution from the crack as it closes and draws in fresh solution as it opens. It is therefore quite common to observe increases in growth rate in water compared with that in air, even though K_{max} in the cycle is less than K_{ISCC}, but effects of frequency, temperature and stress ratio need further investigation. Results in vacuum are generally important, not only to establish scientific mechanisms for growth, but also because vacuum growth models the buried defect, whose detection by NDT is more difficult. A vacuum environment is also needed to establish growth mechanisms at high temperature. Here, creep-fatigue interactions are postulated, but it is necessary to be able to separate effects of high temperature oxidation from those of the deformation mechanisms. Attention might be drawn to some very recent work at Cambridge involving a combination of fatigue and dynamic segregation at high temperature. This is being carried out by Dr. P Bowen & M.B.D.Ellis.

Finally, mention must be made of the growth of cracks, particularly short cracks, at high stress levels, strictly outside the LEFM regime (Taylor and Knott 1981b). Methodologies to treat such growth, using $\Delta\delta$ or ΔJ, are being developed, but it is important to place the work in its engineering context. There are applications, such as the high-temperature disc in a gas-turbine aero-engine, in which defects are small (< 0.15 mm) and in which the stress is high (the bore stress approaching the yield stress). The important feature is that the design life is probably not more than approximately 10^5 cycles. Here, there is a genuine design problem which requires detailed work to quantify crack growth rates in high-stress conditions. In many other cases, engineering design can be based on the lower of two load-ranges (suitably reduced by safety-factors). One is the endurance limit (for 10^7 or 10^8 cycles) or the fatigue limit: the other is calculated from ΔK_{th} (or ΔK_c) and a knowledge of initial defect size and required lifetime. The initial defect size is determined either by NDT or by process control. It is clearly of qualitative and scientific value to appreciate that the actual value of endurance limit is controlled by a defect distribution, but the most efficient method for determining quantitatively the effect of a reduction in defect size on the endurance limit is to test a new set of S-N specimens made from the cleaner material. A further point is that the application of low ΔK LEFM data to natural defects is strictly non-proven. Effects of "plastic-wake" closure have been mentioned above, and it should be noted that results in nickel-base superalloys for short cracks at low ΔK are markedly different from the results for long cracks. There is much research in the area of short crack growth, and it is important that it concentrates on real engineering needs.

The paper began with a summary of Sir Alan Cottrell's comments on our understanding of the science of fatigue. Over the intervening eight-year period, LEFM fatigue-crack growth-rate calculations have been used increasingly in engineering design and to assess the dangers of defects discovered during periodic NDT inspections. Values of da/dN and ΔK_{th} for different materials are now readily available. Even if we still do not fully comprehend separation processes in fatigue at the atomic level, our understanding of the importance of microstructure with respect to growth rate has improved markedly and the present paper has attempted to provide a consistent basis for physical modelling. Further development of our understanding will enable us to provide safer and more efficient designs for engineering structures subjected to cyclic loading and in these endeavours we are simply following the principles laid down over the last thirty years or so by the men to whom this volume is dedicated: Peter Forsyth, Norman Frost and Gerry Smith.

ACKNOWLEDGEMENTS

I wish to thank Prof. R.W.K. Honeycombe F.R.S. and Prof. D. Hull for the provision of research facilities and acknowledge with pleasure the contributions made by Mr. C.H.D. Arbuthnot, Dr. J.E.M. Braid, Dr. G. Clark, Prof. G.G. Garrett, Haicheng Gu, Dr. Y. Higo, Dr. M.N. James, Miss J.M. Kendall, Dr. J.E. King, Dr. V.B. Livesey, Dr. A.C. Pickard, Prof. R.O. Ritchie, Dr. S.M. el Soudani and Dr. D. Taylor. Special thanks are due to Gerry Smith who, over the years, has been a much-respected colleague and a good friend.

REFERENCES

Beevers, C J (1977). Fatigue Crack Growth at Low Stress Intensities, Metal Science, Vol. 11, p. 362

Beevers, C J (1983). Factors Influencing the Rate of Fatigue Crack Growth at Low Stress Intensities proc. Conf. Fracture in the Energy and Transport Systems, ed. I LeMay and S N Monteiro publ. EMAS, Vol. II

Beevers, C J (1985) this volume

Beevers, C J, Knott, J F and Ritchie, R O (1975). Some Considerations of the Influence of Sub-Critical Cleavage Growth During Fatigue Crack Propagation in Steels, Metal Science, Vol. 9, p. 119

Bowen, P (1985) private communication

Brown, L M (1977). Dislocation Substructures and the Initiation of Cracks by Fatigue, Metals Science, Vol. 11, p. 315

Coffin, L F Jnr (1959). Cyclic Straining and Fatigue in Internal Stresses and Fatigue, ed. G M Rossweiler and W L Grube, Elsevier, p. 363

Cottrell, A H (1977). Fracture and Society ICF Interview in Proc. 4th Intl. Cong. on Fracture, Waterloo, ed. D M R Taplin, University of Waterloo Press/Pergamon, Vol. 4, p. 7

Cottrell, A H and Hull, D (1957). Extrusion and Intrusion by Cyclic Slip in Copper, Proc. Roy. Soc. A242, p. 211

El Soudani, S M (1930). Fundamentals of Quantitative Fractography, Ph.D. Thesis, University of Cambridge

Feltner, C E (1965). A Debris Mechanism of Cyclic Strain Hardening for F C C Metals Phil. Mag. 12, p. 1229

Finney, J M and Laird, C (1975). Cyclic Deformation of Copper Single Crystals, Phil. Mag. 31, p. 339

Forman, R G, Kearney, V E and Engle, R M (1967). ASME Jnl. Basic Eng. 89, 459

Forsyth, P J E (1957). Slip-Band Damage and Extrusion, Proc. Roy. Soc., A242, p. 198

Forsyth, P J E and Ryder, D A (1961). Some Results of the Examination of Aluminium Alloy Fracture Faces, Metallurgia 63, p. 117

Frost, N E, Holden J and Phillips, C E (1961). Experimental Studies into the Behaviour of Fatigue Cracks in Cranfield Crack Propagation Symposium paper B2

Garrett, G G and Knott, J F (1975). Crystallographic Fatigue Crack Growth in Aluminium-Copper Alloys, Acta Met. 23, p. 841

Garrett, G G and Knott, J F (1976). On the Influence of Cyclic Hardening and Crack Opening Displacement (COD) on Crack Advance During Fatigue, Met. Trans. 7A, p. 884

Garrett, G G and Knott, J F (1977). On the Effect of Crack Closure on the Rate of Fatigue Crack Propagation, Intl. Jnl. Fracture 13, p. 101

Higo, Y., Pickard, A C and Knott, J F (1981). Effects of Grain-Size and Stacking-Fault Energy on Fatigue-Crack-Propagation Thresholds in Copper-Aluminium Alloys, Metal Science, 15, p. 233

James, M N and Knott J F (1985). Near-Threshold Fatigue-Crack Growth in Air and and Vacuum, Scripta Met. 19, p. 189

James, M N and Knott, J F (1985). An Assessment of Crack Closure and the Extent of the Short Crack Regime in Q1N(HY80) Steel, Fat. and Fract. of Eng. Matls. and Structures, 8, p. 177

James, P L. The Fatigue of Aluminium Alloys Ph.D. Thesis, University of Cambridge 1960

Kendall, J M and Knott, J F (1984a). The Influence of Microstructure and Temperature on Near-Threshold Crack Growth in Air and Vacuum in Fatigue 84, ed. C J Beevers, publ. EMAS, p. 307

Kendall, J M and Knott, J F (1984b). Near-Threshold Fatigue-Crack Growth in Air and Vacuum in Fundamental Questions and Critical Experiments in Fatigue, Dallas 18-23 Oct. 1984, ed. J T Fong to be published (ASTM).

King, J E (1981). Crystallographic Fatigue Crack Growth in Nimonic AP1, Fat. Eng. Matls. and Structures, 4, p. 311

Knott, J F (1973). Fundamentals of Fracture Mechanics, publ. Butterworths (London)

Knott, J F (1980). Micromechanisms of Fibrous Crack Extension in Engineering Alloys Metal Science, 14, p. 327

Knott, J F (1982). Damage Tolerance through the Control of Microstructure in Fracture Mechanics Technology Applied to Material Evaluation and Structure Design, ed. G C Sih, N E Ryan and R Jones. Martinvs Nijhoff 1983, p. 19.

52

Knott, J F and Pickard, A C (1977). Effects of Overloads on Fatigue Crack Propagation: Aluminium Alloys, Metal Science, 11, p. 399

Laird, C (1962). Studies of High-Strain Fatigue, Ph.D. Thesis, University of Cambridge

Laird, C and Smith, G C (1962). Crack Propagation in High Stress Fatigue, Phil. Mag., 7, p. 847

Lardner, R W (1967). A Dislocation Model for Fatigue Crack Growth in Metals, Phil. Mag. 17, p. 71

Lio, H W (1961). Trans. ASME Jnl. Basic Eng. 83, p. 23

McEvily, A J (1977). Current Aspects of Fatigue, Metal Science, 11, p. 274

McEvily, A J and Groeger, J (1977). On the Threshold for Fatigue Crack Growth, Proc. 4th Intl. Cong. on Fracture, ed. D.M.R. Taplin, publ. Univ. of Waterloo Press/Pergamon II, p. 1293

Manson, S (1954). U.S. Nat. Advis. Ctte, Aero Tech. Note (2933)

Paris, P and Erdogan, F (1963). Trans. ASME Jnl. Basic Eng. 85, p 528

Pickard, A C (1977). A Study of Fatigue Crack Propagation in Metals, Ph.D. Thesis, University of Cambridge

Pickard, A C, Ritchie, R O and Knott, J F (1975). A Study of Fatigue Crack Propagation in a Type 316 Stainless Steel Weldment, Metals Technology, p. 253

Pickard, A C, Ritchie, R O and Knott, J F (1976). Fracture Toughness on Fatigue Crack Propagation Studies in a Complex Aluminium Bronze, Proc. 4th Intl. Conf. on Strength of Metals and Alloys, Nancy, Vol. 2, p. 473

Ritchie, R O (1981). Environmental Effects on Near-Threshold Fatigue Crack Propagation in Steels: a Re-Assessment in Fatigue Thresholds, Ed. J. Bäcklund, A.F. Blom and C.J. Beevers, EMAS, 1982, p. 503

Ritchie, R O and Knott, J F (1973). Mechanisms of Fatigue Crack Growth in Low Alloy Steel, Acta Met. 21, p. 639

Ritchie, R O and Knott, J F (1974a). Microcleavage Cracking during Fatigue Crack Propagation in Low Strength Steel, Mater. Sci. and Eng. 14, p. 7

Ritchie, R O and Knott, J F (1974b). On the Influence of High Austenitizing Temperatures and Overheating on Fracture and Fatigue Crack Propagation in a Low Alloy Steel, Met. Trans. 5, p. 782

Suresh, S, Zavinski, G F and Ritchie, R O (1981). Oxide-Induced Crack Closure: An Explanation for Near-Threshold Corrosion Fatigue Crack Growth Behaviour, Met. Trans., 12A, p. =435

Taylor, D and Knott, J F (1981). Fatigue-Crack Propagation Behaviour of Short Cracks: The Effect of Microstructure, Fat. Eng. Matls. and Structures, 4, p. 147

Taylor, D and Knott, J F (1982). Growth of Fatigue Cracks from Casting Defects in Nickel-Aluminium Bronze, Metals Technology, 9, p. 221

Thompson, N and Wadsworth, N J (1958). Metal Fatigue, Advances in Physics, 7, p. 72

Tomkins, B (1968). Fatigue Crack Propagation - An Analysis, Phil. Mag. 18, p.1041

Weertman, J (1969). Theory of Rate of Growth of Fatigue Cracks and Combined Static and Cyclic Stresses, Int. Jnl. Fract. Mech., 5, p. 13

Metallurgical Aspects of
Fatigue Crack Growth

T. C. LINDLEY and K. J. NIX

*Technology Planning and Research Division, Central Electricity Generating Board,
Central Electricity Research Laboratories, Leatherhead, UK*

ABSTRACT

Defects are introduced into metals by processing, fabrication, environmental attack
and service stressing and these processes are briefly reviewed. The metallurgical
factors affecting the development of short cracks are discussed and comparisons made
with the growth of long cracks.

Emphasis is placed on the role of inclusions in the formation and growth of fatigue
cracks.

Finally, current metallurgical research aimed at improving fatigue properties is
briefly discussed.

KEYWORDS

Defect formation; microcrack growth; macrocrack growth; microstructure, inclusions;
improved fatigue properties.

INTRODUCTION

This paper discusses the ways in which metallurgical factors can affect the development of a growing fatigue crack. The various ways by which defects are introduced
into materials are briefly considered followed by more detailed appraisal of the
metallurgical factors which can influence the growth of both small and large fatigue
cracks. Finally, alloy development research aimed at improving fatigue performance
is briefly discussed.

FORMATION OF MATERIAL DEFECTS

Defects can be introduced into components by various routes including processing,
fabrication, environmental attack and service stressing.

Processing

Components may be described after the three main processing routes: wrought (or cast
and wrought), cast and powder formed. For wrought products, both composition and
degree of deoxidation in the initial casting determine the degree of porosity,
segregation and distribution of inclusions prior to working. Subsequent working can

FCG—C

53

modify the shape and distribution of these defects as well as introducing grain texture and anisotropy of mechanical properties. The morphology of these defects determines the degree to which they behave as cracks. For example, sulphide inclusions in structural steels are plastic at hot working temperatures and are rolled out into thin, crack-like platelets up to 1 mm in length. Such defects can then act as either crack initiators or crack arrestors depending upon their orientation. In steels, the high temperature plasticity of inclusions can be varied by changing the oxygen content or by modification e.g. by addition of calcium or rare earth elements. If the steel has very low sulphur content, then oxide or silicate inclusions will predominate, rather than sulphides. On cooling, differential thermal contraction causes residual stresses to develop at inclusions, sulphides being left in tension whereas oxides and silicates are left in compression. The effect of morphology and local residual stress pattern on the initiation and early growth of fatigue cracks at sulphides and oxides in steels has recently been studied by Thomason (1985).

Final defect sizes in wrought products depend upon both material quality and the precise working process and can commonly range from a few microns to about one millimetre.

Cast to shape products can contain a variety of defects including shrinkage cavities, gas porosity, slag entrapment and solid non-metallic de-oxidation products. Casting alloys sometimes have microstructures comprising a low density phase distributed in a metallic matrix. For example, grey cast iron and aluminium-silicon alloys in the unmodified condition contain distributions of graphite flakes and silicon needles respectively. These second phases provide easy crack initiation sites. Defect sizes in castings are commonly in the range 0.1 up to several millimetres. The recognition of the presence of these defects in castings has led to the use of higher safety factors than for wrought material.

Certain alloys cannot easily be made by the conventional cast and wrought processing route which can give segregation problems. For this reason, complex nickel base alloys are made by a powder compaction route involving hot isostatic pressing and forging. This process can introduce foreign particles (frequently of refractory material) of a size comparable with the initial powder size i.e. 0.1-0.2 mm into the final product.

Defects up to about 1 mm in size can therefore be introduced into components by a variety of means. The likelihood of such defects initiating cracking during service will depend upon their detailed morphology, location (surface breaking or buried) and orientation to the applied stress field.

Fabrication

Of the various fabrication routes, fusion welding is likely to produce the greatest variety of defects in metal-metal joints. Soldering, brazing and various forms of pressure or friction welding can also introduce defects through the formation of brittle intermetallic compounds or incompletely bonded interfaces.

With fusion welding, slag-entrapment or lack of fusion may produce significant defects in the weld metal. In the absence of adequate liquid metal feed from a 'matched' welding consumable, solidification cracking can occur, exacerbated by the residual stresses generated as the cast weld metal cools. Hydrogen uptake may result in cracking in the heat affected zone whilst lamellar tearing is promoted when elongated inclusions are subject to transverse stresses. Finally, stress relief (reheat) cracking can occur in some quench and tempered steels, particularly in the presence of high levels of trace impurities.

A wide range of defect sizes can be produced by welding processes but, with care, they can be restricted to the sub-millimetre range.

Removal of the surface layer by grinding and machining can also introduce small defects associated with complex residual stress fields, particularly in the harder alloys. For example, grinding can introduce 'white layers' of very hard, untempered martensite. Machining itself can give rise to a wide variety of surface finishes with grooves typically 1 µm for a fine turned finish and 10-20 µm for a rough finish. Beneficial compressive residual stresses from peening processes can be counterbalanced if rough surfaces are produced and the incipient cracks (size typically 5-20 µm from surface undulations) can promote crevice corrosion.

Heat-treatment of alloy steels can cause quench-cracks due to differential transformation strains. Decarburization of carbon steels promotes easy surface deformation which may reduce fatigue crack initiation life.

Hard surfaces produced by heat-treatment (and chemical diffusion) during carburizing and nitriding etc. can also be prone to cracking. Although such hard layers can retard initiation and early growth in load-controlled fatigue, plastic strain control can introduce an array of small cracks to the depth of the hardened layer, in the first tensile half cycle. Typical defect sizes are 1-2 mm for a carburized layer and 0.5 mm for nitriding.

Environmental Attack

If the service environment is chemically aggressive, then the component surface may suffer local attack by intergranular corrosion or pitting. Corrosion pits commonly form by anodic dissolution at certain favoured sites. These include (a) slip steps where the protective oxide film is locally ruptured. (A review of pit formation by environmental attack at persistent slip bands has recently been made by Yan, Farrington and Laird, (1984)) (b) grain boundaries particularly if temper embrittled (alloy steels) or having precipitate-free zones (aluminium alloys) (c) chromium depleted zones in stainless steels (d) non-metallic inclusions e.g. MnS particles in steels (e) multiphase alloys where one of the phases is preferentially attacked. Anodizing pits can form in aluminium alloys due to dissolution of intermetallic particles, Forsyth (1979).

Pits can develop in a variety of geometrical shapes and they are frequently sub-millimetre in size. Pits are local stress raisers and are favoured sites for fatigue crack initiation. It has recently been shown that they can significantly reduce fatigue strength in a variety of materials e.g. NiCrMoV alloy steels, Lindley, McIntyre and Trant (1982), austenitic stainless steels, Amzallag (1977), Marshall (1985) and aluminium alloys Weston and Wilson (1981).

Intergranular corrosion cracks are commonly a few grains deep.

Service Stressing

Defects in the form of small fatigue cracks can be introduced by plastic deformation processes arising from service loading. These defects are normally confined to the material surface where lack of constraint aids crack formation. In a poly-crystalline material, plastic deformation is triggered in certain favourably oriented surface grains to form so-called 'persistent slip bands'. Intrusions (1-10 μm in size) are created where the persistent slip band intersects the free surface. Since such intrusions are stress raisers (on a microscopic scale), they will develop by promoting more plastic deformation. The initiation and early growth of such defects will depend upon the material's plastic deformation characteristics i.e. whether slip is planar or wavy, the cyclic hardening or softening response, the ability to strain-age (a process which may be responsible for the phenomenon of non-propagating cracks, Knott (1985)).

However, rather than introducing small 'natural' defects, the role of plastic deformation in promoting crack development at inclusions is likely to be of greater relevance.

THE DIFFERENT STAGES IN FATIGUE LIFE

The degree of perfection of the starting material will determine the route by which a significant fatigue crack is formed. In general the easiest route will be chosen. For example, although the persistent slip band mechanism of crack formation might be necessary in pure metals, the presence of inclusions (or surface scratches) in commercial materials will greatly speed up the crack initiation process. This will be even more the case in weldments where defects up to 1 mm in size can be present before the component enters service. In Fig. 1 after Schijve (1979), it can be seen that for both the polished surface (pure metal) case and the commercial material (inclusion) case, the formation of a small crack ≈100 μm in size can take up 60-80% of fatigue life. It is for this reason that there is so much current interest in the growth behaviour of short cracks. Since these microcracks will be similar size to the material grain size, it might be expected that microstructure would strongly influence the development of a small crack, a topic now considered.

GROWTH OF MICROCRACKS

The earliest direct observations of the mechanisms of microcrack growth were made by Forsyth and his co-workers (1963). He showed that the initiation and early growth of microcracks can occur by the activation of a single slip system (Stage I crack growth) in a favourably oriented surface grain (Fig. 2). In peak-aged high purity AlZnMg alloys, stage I cracking was further enhanced by the formation of 'softened' persistent slip bands due to dislocation cutting of precipitates and

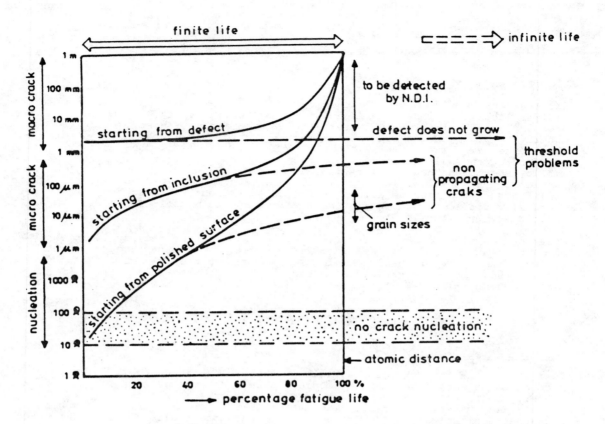

Fig. 1. Different phases of fatigue crack development (after Schijve (1984))

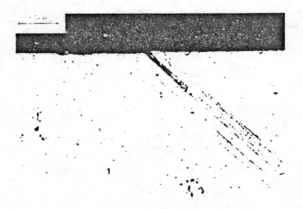

Fig. 2. Stage 1 fatigue crack associated with persistent slip band in Al-Zn-Mg alloy (after Forsyth and Stubbington (1962))

their subsequent dissolution. If the alloys were overaged to prevent precipitate-cutting, transgranular Stage I growth was suppressed and cracking occurred (Fig. 3) along the precipitate free zones adjacent to favourably oriented surface grain boundaries (Forsyth and Stubbington (1962)). The transition from Stage I to a crack growth mechanism involving the activation of multiple slip systems (Fig. 4) at the crack tip (Stage II growth) can occur at or near the first grain boundary encountered by the crack. As might be expected, grain boundaries can have a significant effect on the growth of microcracks, which will be discussed later.

The growth of 'natural' surface initiated short cracks in commercial aluminium alloys has been investigated by Morris, Buck and Marcus (1976); Kung and Fine (1979)

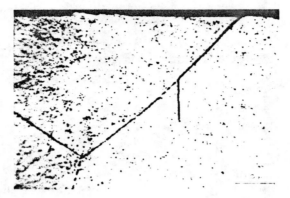

Fig. 3.   Stage 1 cracking associated with precipitate
free zone at a grain boundary in Al-Zn-Mg
alloy (after Forsyth and Stubbington (1962)

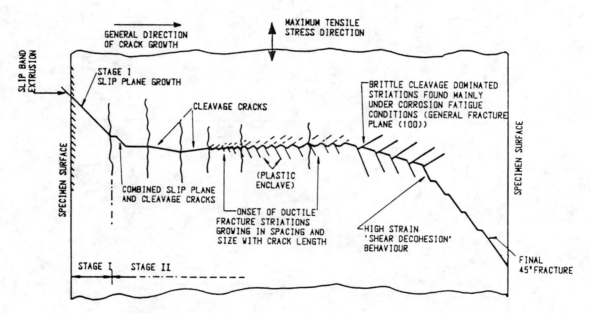

Fig. 4.   Various modes of fatigue crack growth in strong
aluminium alloys (after Forsyth (1963))

and Lankford (1983). In some cases, microcracks initiated at inclusions and the
Stage I period of crack growth was eliminated. This tendency toward inclusion
initiation rather than slip band (Stage I) cracking was found to depend on stress
level and inclusion content (Kung and Fine (1979)). Slip band cracking was promoted by high stresses and higher alloy purity.

The observation that short cracks can apparently grow at more rapid rates than those
predicted by linear elastic fracture mechanics (LEFM), particularly in the near-
threshold growth regime, has attracted considerable attention. Earlier work by
Pearson on fatigue crack initiation and growth from inclusions in age-hardened
commercial aluminium alloys (BS L 65 and DTD 5050) is shown in Fig. 5. Pearson (1975)
concluded that cracks of size approximating to the average grain-size, grew several
times faster than long cracks at nominally identical ΔK values. There has been much
discussion of Pearson's observations. It has been argued that the calculation of
ΔK for a short crack growing from the initiating inclusion could be in error
(Schijve (1984)). For example, if initiation occurs sub-surface with subsequent
breakthrough to the surface, a considerable elevation in ΔK is possible over that
calculated from surface observations (Fig. 6). The same reservation can be made of
De Lange's (1964) work, about ten years earlier than Pearson's (1975). Although it
has proved convenient to use ΔK to characterize the growth of short cracks, its

58

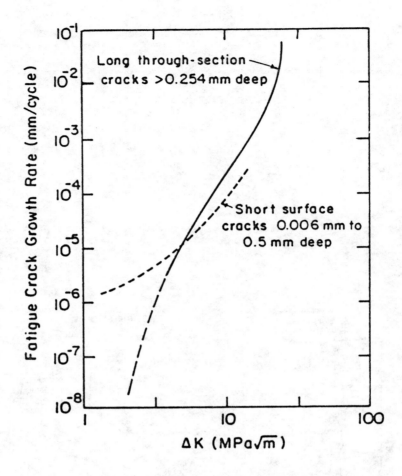

Fig. 5. Comparison of growth rates for short and long
cracks as a function of ΔK in aluminium alloys
(after Pearson (1975))

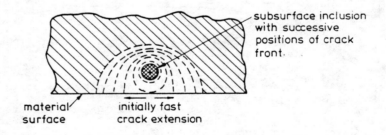

Fig. 6. Sub-surface crack initiation suggesting an
initial fast microcrack growth (after
Schijve (1984))

universal application can be viewed with some scepticism. Despite the above quali-
fications, recent work on the growth of 'natural' short cracks, notably by Lankford
(1983) and Morris, Buck and Marcus (1976), has demonstrated the veracity of high
crack growth rates and low threshold values and this work will now be considered in
some detail.

Lankford (1983 and Morris, Buck and Marcus (1976) studied cracks initiated from
inclusions in aluminium alloys and found that microcracks growing in the first

(surface) grain, grew more rapidly than long cracks and that these short cracks could propagate at ΔK levels below the long crack threshold, ΔK_{th}. Lankford subjected a tensile geometry specimen to direct stress, rather than use the more complex geometries of rotating bend and cantilever bend employed by De Lange (1964) and Pearson (1975) respectively. As the microcrack approached the first grain boundary, crack growth retardation or even arrest could occur, depending upon the degree of orientation mismatch across the boundary. Figure 7 demonstrates that minimum growth

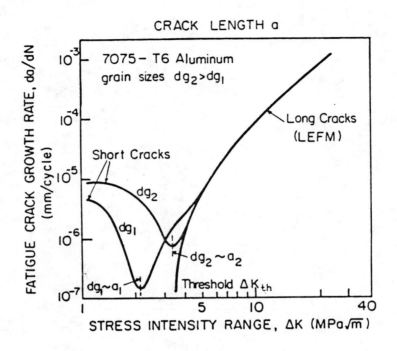

Fig. 7. Effect of grain-size (dg) on growth of short and
long cracks in aluminium alloys. Minima in
growth rates occurs when crack size approximates
to grain-size (after Lankford (1983))

rate occurs when the crack-size is approximately equal to the grain-size. The retardation effect may result from the difficulty of nucleating microplasticity in the new grain (Morris, James and Buck (1981), Lankford (1983)). Alternatively, crack deflection at the grain boundary may cause increased crack closure (Suresh (1983)). Microcracks which experience little orientation mismatch at the grain boundary propagate smoothly, eventually achieving growth rates which coincide with long crack data (Fig. 8).

The crack growth retarding effect of grain boundaries demonstrates that the microcrack is smaller than the size scale at which the material may be considered homogeneous and strictly should not be characterized using continuum mechanics. In the LEFM regime, the fatigue crack driving force is the crack tip opening and closing which is achieved by plastic deformation processes at the crack tip itself. By contrast, microcracks initiate and grow in favourably oriented grains which are already experiencing microplasticity. An example of such a microcrack growing in a local plastic strain is the case of a Stage I crack growing in a pre-existing persistent slip band.

It is clear from the foregoing discussion that grain boundaries can profoundly influence the growth of short cracks whereas they generally have little effect on long crack growth (see Section 5). Cook and his co-workers (1981) studied the crack tip deformation associated with the growth of both short and long cracks in the fine-grained nickel based superalloy IN-100. They found that the macroscopic deformation field at the crack tip was the same for both short and long cracks and was controlled by stress intensity factor. However, the fatigue crack tip was found to interact strongly with grain boundaries. Consequently, local crack growth rates could not be related to fracture mechanics parameters. Since the tip of a long crack intersects

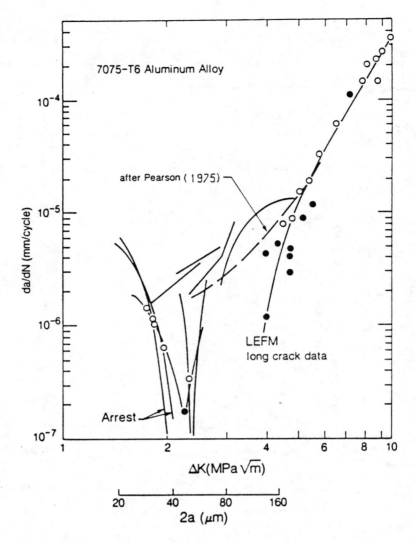

Fig. 8. Comparison of growth of short and long cracks
in aluminium alloys. Large grain boundary
mismatch causes crack arrest whereas small
mismatch causes no transient (after Lankford
(1983))

many grains of different orientations, any local effects are averaged out and the
macroscopic growth rate of the long crack can be related to ΔK.

In heavily textured aluminium alloy, Gregory, Gysler and Luetjering (1984) have shown
that the grain boundary retardation effect is highly dependent on test orientation
due to the misorientation influence.

Taylor and Knott (1981) have attempted to relate the 'anomalous' short crack growth
behaviour to some microstructural unit such as grain size, using the above argument
that until the crack tip plastic zone includes a large number of grains crack growth
cannot be treated using continuum mechanics. They found that for a number of
materials that a_2 (Fig. 9) was approximately equal to 10 d where d is some microstructural unit size, usually the grain-size. The parameter a_2 is the dimension
above which crack size has no influence on threshold ΔK_{th} (Fig. 9).

The high fatigue crack growth thresholds (long cracks) observed in β processed
α-titanium alloys such as IMI 685 have been related to extensive crack closure
resulting from a highly tortuous crack path, Halliday and Beevers (1981). This
'crack path tortuosity' results from preferred crystallographic growth on the basal
plane in this close packed hexagonal metal. The pronounced 'roughness induced'
crack closure reduces the crack driving force $\Delta K_{effective}$ and therefore the growth
rate. Experimental measurements of crack closure levels by various workers

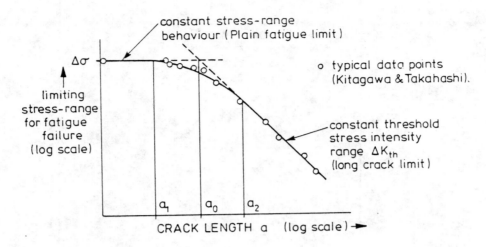

Fig. 9. The limiting stress range for fatigue failure
as a function of crack length (after Taylor and
Knott (1981))

(Morris, James and Buck (1981); Breat, Mudry and Pineau (1983); Minakawa and McEvily
(1981)) have shown that short cracks show less closure than long cracks. This
mechanical factor accounts, at least in part, for the faster growth rates of short
cracks. However, crack closure arguments alone cannot account for the observed
higher growth rates of short cracks in IMI 685 alloy (Brown and Hicks (1983)),
microstructural factors also being important. Alloy IMI 685 can exhibit a very large
grain-size (e.g. prior β grains $\approx$5 mm diameter, α platelet size $\approx$1 mm in the study
of Brown and Hicks (1983)). Hence the 'short crack regime' in this alloy will be
more extensive than usual. Brown and Hicks (1983) measured enhanced growth rates
for cracks up to 3.5 mm deep, spanning the crack size growth regime of great practical
significance. Short crack growth rates were found to increase with micromechanism of
growth in the order (i) non-crystallographic (ii) colony boundary separation and
(iii) crystallographic growth on the basal plane (Fig. 10(a) and (b)). Whilst the

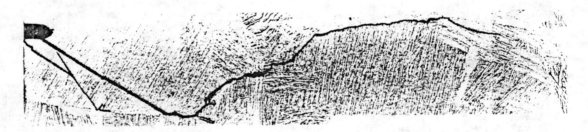

Fig. 10(a). Micrograph showing crystallographic and colony
boundary separation in crack growth from a
notch (after Brown and Hicks (1983))

rapid crystallographic growth mechanism can still occur with long cracks (Ward-Close
and Beevers (1980)) within certain favourably oriented grains, the activation in
adjacent grains of the slower mechanisms retards average growth. For a short crack
contained within one or two grains, however, the more rapid growth mechanism can
prevail. For crystallographic growth, Brown and Hicks (1983) postulated that slip
is concentrated into a narrow band ahead of the crack. This allows the development
of high shear strains at the crack tip, resulting in elevated growth rates.

In assessing the effect of grain-size on growth behaviour, it is informative to
compare fatigue crack growth rates of short and long cracks in single crystals as
well as polycrystals. Hicks and Brown (1984) have recently made such a comparison
in the nickel base superalloy SRR 99 (Fig. 11). Little effect of crack length on
growth rate was observed and the growth of short cracks was adequately characterized

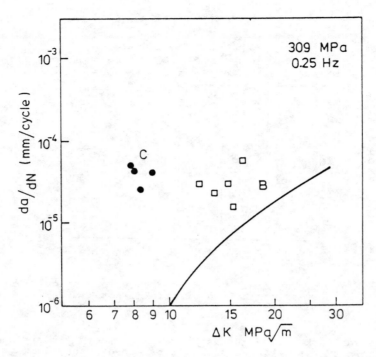

Fig. 10(b). Fatigue crack growth rates corresponding
to the crystallographic (C) and colony
boundary separation modes (B). Solid
line represents long crack behaviour
(after Brown and Hicks (1983))

using ΔK. It is important to note that both the short and long cracks in this study
were 'short' in relation to the grain-size. Even in this case LEFM modelling will
still be inadequate at very small crack sizes when the scale of plasticity is large
compared to the crack size.

Short crack growth behaviour can be conveniently separated into the different fatigue
regimes viz (1) low stress/high cycle in plain specimens (2) low stress/high cycle
in notched specimens (3) high stress/low cycle in plain specimens (4) high stress/
low cycle in notched specimens. In these different regimes, the mechanics of growth
often controls fatigue behaviour. However, metallurgical factors such as strength
level (and grain-size) can also be of great significance.

In relation to the high cycle fatigue regime, the well known plot of the form shown
in Fig. 9 of Kitagawa and Takahashi (1976) showing the transition as crack-size
decreases from LEFM controlled (ΔK) growth to stress controlled behaviour as the
fatigue limit is approached. The importance of this plot is that it gives the range
of validity of LEFM and Fig. 12 shows that the transition crack size is material
dependent. High strength and mild steels show transition crack sizes of ~ 10 μm and
~ 500 μm respectively.

In moving on to the growth of long cracks and to provide a link between short and
long crack behaviour, the mechanistic and materials aspects which possibly cause the
differences between these are summarized in Figs. 13 and 14 (Schijve 1984). In
addition to promoting microcrack formation, inclusions can have a retarding effect
on the growth of macrocracks (Forsyth and Bowen (1981)).

The pronounced anisotropy in fatigue behaviour due to inclusions in steels will be
discussed in a later section.

GROWTH OF MACROCRACKS

The well-known sigmoidal plot (Lindley, Richards and Ritchie (1976)) of da/dN versus
ΔK is shown in Fig. 15 and indicates the factors which can affect crack growth in
steels in the different fatigue regimes. It will be noted that microstructure is
particularly important in the near-threshold regime (Region A) and the static modes

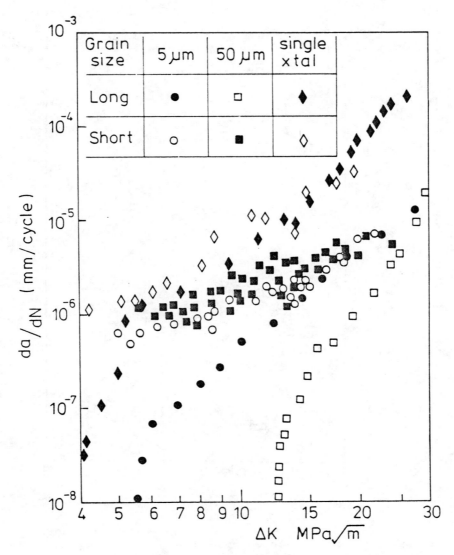

Fig. 11. The effect of microstructure on fatigue crack
growth rates in nickel base alloys (after
Hicks and Brown (1984))

regime (Region C) although in the latter case these faster growth rates usually have
little effect on overall fatigue life.

Near Threshold Growth Regime

In the near-threshold regime where the cyclic plastic zone is of the order of the
grain size, growth becomes sensitive to microstructure. The metallurgical factors
affecting near-threshold growth include strength level and grain size. Figure 16
shows that at a given value of stress ratio R, threshold ΔK_{th} decreases with
increasing strength level in steels. In essentially single phase microstructures
such as aluminium, titanium, ferritic low carbon iron and austenitic stainless steel,
long crack threshold ΔK_{th} increases with increasing grain size. Figure 17 shows
ΔK_{th} increasing with increasing grain size in 316 stainless steel in the work of
Priddle (1978).

In the more complex bainitic and martensitic steels, Ritchie and his co-workers (1979)
have shown that ΔK_{th} is little affected by prior austenite grain size, although it
might be argued that in these steels, carbide cell size is the more relevant microstructural parameter. Figure 18 shows the contrasting behaviour between the low
carbon iron and high strength steels with respect to grain size.

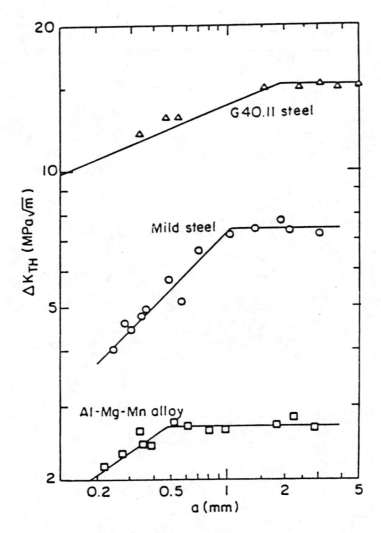

Fig. 12. Variation of threshold stress intensity
factor ΔK_{th} at short crack sizes for
different materials (after Suresh and
Ritchie (1984))

Fatigue strength (based on S-N data) and fatigue threshold ΔK_{th} invariably show
opposite trends when plotted against strength level (Fig. 19). Refining the grain
size (and concurrently increasing the material strength level) will give increased
resistance to crack initiation but decreased resistance to crack growth.

Several interesting research projects are currently in progress aimed at improving
high cycle fatigue properties by microstructural control. Byrne and his co-workers
(1981) have produced a high strength pearlitic steel with good fatigue properties
by a thermo-mechanical heat-treatment (TMT) which consists of heavy cold-rolling
followed by rapid annealing. The resulting microstructure consists of aligned
cementite plates in a soft ferrite matrix and gives good fatigue strength and high
threshold value ΔK_{th}.

In other alloy systems (AISI 1018 and 2¼CrlMo low alloy steels) with duplex ferrite/
martensite microstructures and where the ferrite is encapsulated by a continuous
martensite phase, it is possible to achieve increasing threshold ΔK_{th} level with
increasing strength level. Inter-critical annealing of duplex alloy steels (Dutta,
Suresh and Ritchie (1983)) can cause elevation of ΔK_{th} to a value as high as
$\sim$ 20 MPa m$^{\frac{1}{2}}$ at low R. Hopefully, in the future, these features will be exploited in
design against fatigue. For commercial alloys in the meantime, it is necessary to
know from the outset if design is to be based on resistance to crack initiation or
crack growth, in order to achieve optimum fatigue performance.

DIFFERENCE BETWEEN THE GROWTH OF

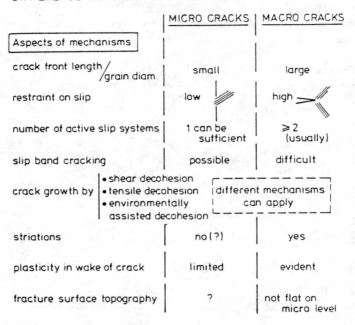

Fig. 13. Differences between the growth of
short and long cracks (after
Schijve (1984))

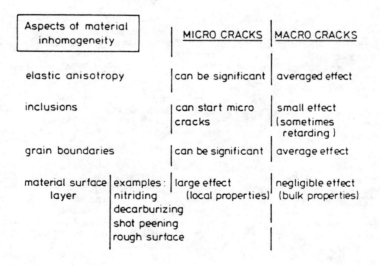

Fig. 14. Effect of material inhomogeneity on
growth of short and long cracks
(after Schijve (1984))

Static Modes Fatigue Regime

The tendency toward higher growth rates in Region C of the sigmoidal plot (Fig. 15)
of da/dN versus ΔK stems from the introduction of growth mechanisms involving
static modes i.e. intergranular, micro-cleavage and void coalescence (Lindley,
Richards and Ritchie (1976). Since static modes are triggered as K_{max} of the
fatigue cycles approaches fracture toughness value K_{1c} (or as σ_{max} approaches a
plastic collapse value), there is a large effect of microstructure on crack growth.
Metallurgical factors which cause reductions in fracture toughness will promote

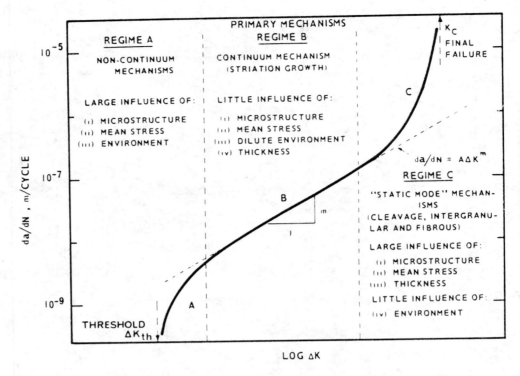

Fig. 15. Sigmoidal variation of fatigue crack growth rate of long cracks as a function of ΔK (after Lindley, Richards and Ritchie (1976)

Region C in Fig. 15. For example, Fig. 20 shows the large effect of temper embrittlement on fatigue crack growth in the low alloy steel En 30A. In the unembrittled condition, crack growth occurs by striation formation and is insensitive to mean stress. By contrast, temper embrittlement promotes a static mode in the form of intergranular cracking and the increased growth rates are accompanied by a large dependence on mean stress (Fig. 20).

In many situations, Regime C may be considered to be of minor practical importance since fatigue crack growth rates are so high and remaining crack propagation life correspondingly low. However, it can be important where large in-service loads occur infrequently, as is the case for certain operational transients in pressure vessels.

Recent CEGB work has been concerned with anisotropy of both S-N and fatigue crack growth properties in rotor and pressure vessel steels. This anisotropy results from inclusions in the steels. Figure 21 shows fatigue crack growth data for an older generation pressure vessel steel having high inclusion content. Fatigue cracks lying in the rolling plane (SL orientation) grow much faster for ΔK > 15 MPa m^½ than cracks growing through the plate in the thickness direction (TL orientation). In the TL orientation, crack growth occurs by the striation mechanism. By contrast, the faster growth rate in the SL orientation is due to the introduction of static modes (in the form of microvoid coalescence) and de-lamination at stringers of non-metallic (Fig. 22) inclusions in the rolling plane. The introduction of these static modes is a direct consequence of the low toughness of the SL orientation due to the inclusion density and morphology. Fortunately, the latest generation pressure vessel steels with carefully controlled inclusion content do not show significant anistropy.

Finally, mention must be made of the exciting development of the aluminium-lithium alloys for use in aircraft structures. An alloy based on AlLiCuMgZr is of particular interest and shows a reduction in density of 10% and increased elastic modulus of 8% compared to the conventional 2014 A aluminium. The zirconium is added to prevent recrystallization and retard grain growth. Copper and magnesium are jointly added to form Al_2CuMg precipitates which disperse slip which is otherwise planar in binary Al-Li. This change in the nature of slip gives improved toughness without loss in ductility. Fatigue crack growth rates in this alloy have been studied by Harris and his co-workers (1984). Figure 23 compares growth rates in AlLiCuMgZr (peak aged condition) with AlCuMg of the same strength level. The lithium containing

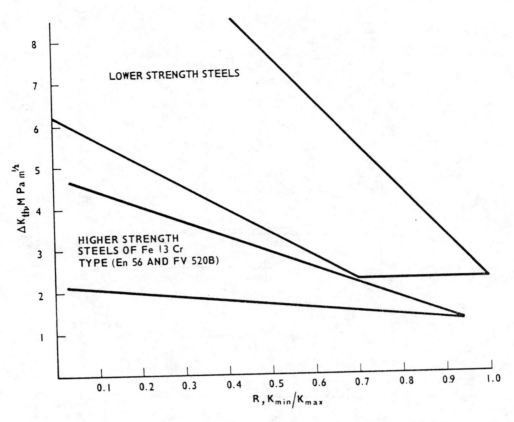

Fig. 16. Threshold ΔK_{th} values as a function of stress ratio R (after Lindley (1985))

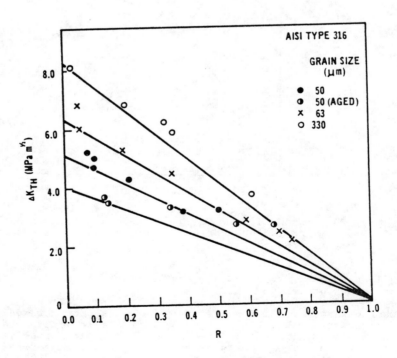

Fig. 17. Effect of grain-size on threshold ΔK_{th} in AISI 316 stainless steel at different values of stress ratio R (after Priddle (1978))

68

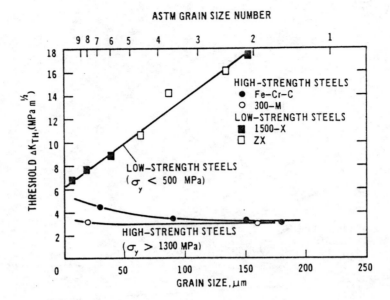

Fig. 18. Variation of threshold ΔK_{th} with grain-size for low and high strength steels at R = 0.05 (after Ritchie (1979)

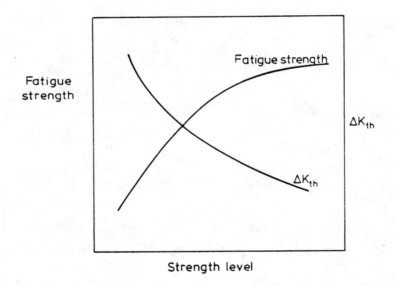

Fig. 19. Schematic of fatigue strength (based on S-N data) and fatigue threshold ΔK_{th} as a function of strength level

alloy shows lower growth rate due, at least in part, to a lower modulus. Figure 24 shows fatigue crack growth in AlLiCuMgZr in various conditions of ageing. Under-ageing gives a low growth rate and a high threshold ΔK_{th} of 5 MPa m$^{\frac{1}{2}}$, compared with AlCuMg in the peak-aged condition. If fatigue crack growth rate is plotted against $\Delta K/E$, then it becomes apparent that the low growth rate for the under-aged condition cannot be explained in terms of modulus alone. An explanation lies with the mechanism of crack growth. The underaged condition shows faceted fracture arising from intense localized shear and cracking on {111} planes (Fig. 25). The zig-zag fracture path thereby produced promotes roughness-induced crack closure which in turn reduces $\Delta K_{effective}$, the crack driving force. By contrast, the overaged condition shows relatively straight fracture paths and less crack closure.

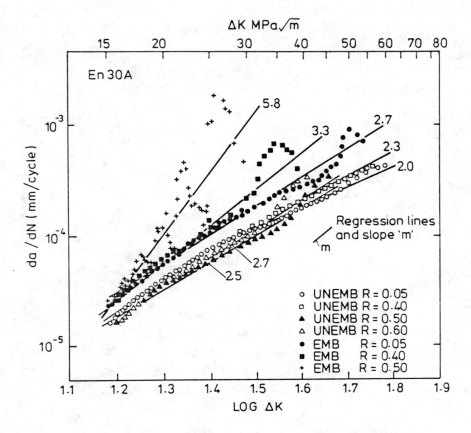

Fig. 20. Variation of fatigue crack growth rate with ΔK in
unembrittled and embrittled En30A steel (numbers
indicate slope m of regression lines drawn
through data points) (after Lindley, Richards and
Ritchie (1976))

CONCLUDING REMARKS

Processing, fabrication, environmental attack and service stressing can each promote
the formation of small defects in components.

Microstructure can strongly influence the development of a short crack. For long
cracks, metallurgical factors are particularly important in the near-threshold and
static modes regimes.

Fatigue properties can be improved by changes in alloy chemistry and by thermo-
mechanical heat-treatment.

ACKNOWLEDGEMENTS

The authors are grateful to their colleagues, particularly Dr J.F. Knott and
Dr B. Tomkins for many valuable discussions within a Royal Society Working Group
considering sub-critical crack growth by fatigue and stress corrosion.

This work was carried out at the Central Electricity Research Laboratories and is
published by permission of the Central Electricity Generating Board.

REFERENCES

Amzallag, C., Rabbe, P., Lieurade, H.P. and Truchon, M., (1977). Conf. on the
Influence of Environment on Fatigue, I. Mech. Eng., London, 638-646.

Breat, J.L., Mudry, F. and Pineau, A., (1983). Fat. Eng. Mats. and Strucs., 6,
349-358.

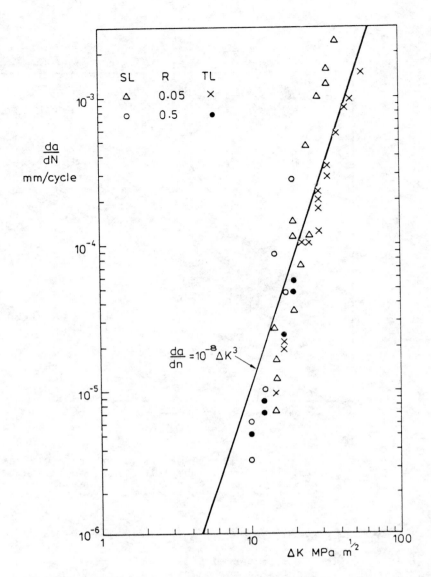

Fig. 21. Fatigue crack growth for TL and SL orientations in rolled steel plate. Line gives upper bound rate for growth by striation formation

Brown, C.W. and Hicks, M.A., (1983). Fat. Eng. Mats. and Strucs., 6, 67-76.

Cook, T.S., Lankford, J., and Sheldon, G.P., (1981). Fat. of Eng. Mats. and Struc., 3, 219-220.

De Lange, R.G., (1964). Trans. Met. Soc. AIME, 230, 644-649.

Dutta, V.B., Suresh, S. and Ritchie, R.O. (1983). Met. Trans. 15A, 1193-1207.

Forsyth, P.J.E., (1963), Acta. Met. 11, 703-707.

Forsyth, P.J.E. and Stubbington, C.A., (1962) RAE Technical Note Met/Phy 344.

Forsyth, P.J.E. (1979). Proceeding of Seeco 79, SEE Fatigue Conference, 9 to 11 May 1979, London 45-66.

Forsyth, P.J.E. and Bowen, A.W. (1981). Int. J. of Fatigue 3, 17-25.

Gregory, J.K., Gysler, A. and Luetjering, G., (1984). Proc. of Fatigue 84, 3 to 7 Sept. 1984, Birmingham University, England, 847-856, Publishers EMAS.

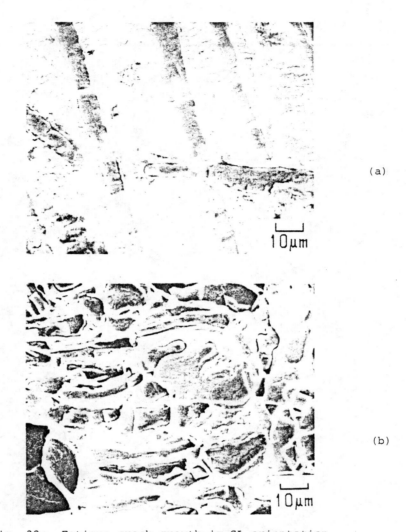

(a)

(b)

Fig. 22.   Fatigue crack growth in SL orientation
(a) delamination of MnS particles at
low ΔK
(b) fracture of MnS particles and void
formation at high ΔK

Halliday, M.D. and Beevers, C.J. (1981). J. of Testing and Evaluation, 9, 195-201.

Harris, S.J., B. Noble and K. Dinsdale (1984), Proc. Fatigue 84, 3 to 7 Sept. 1984.
  Birmingham University, England, 361-370, Publishers EMAS.

Hicks, M.A. and C.W. Brown (1984). Proc. of Fatigue 84, 3 to 7 Sept. 1984,
  Birmingham University, England, 1337-1348, Publishers EMAS.

Journeaux, G.E., Martin, J.W. and Talbot, D.E.G. (1979). Conf. on Mechanisms of
  Environment Sensitive Cracking of Materials, Guildford, England. Metals Society,
  322-333.

Kao Po-We and Byrne, J.G. (1981). Int. Conf. on Fat. Thresholds, Stockholm June,
  Paper No. 22, 313-328, Publishers EMAS.

Kitagawa, H. and Takahashi, S., (1976). Proc. 2nd Int. Conf. Mech. Behav. of Mats.
  Boston, Mass., 627-631.

Knott, J.F., 1985. Private Communication.

Kung, C.Y. and Fine, M.E., (1979). Met. Trans. A, 10A, 603-611.

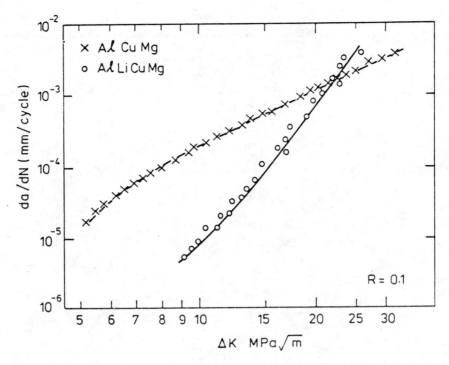

Fig. 23. Fatigue crack propagation rates in peak aged AlLiCuMgZr and AlCuMg of the same strength level OA, PA and UA represent overaged, peak-aged and underaged conditions respectively (after Harris, Noble and Dinsdale (1984))

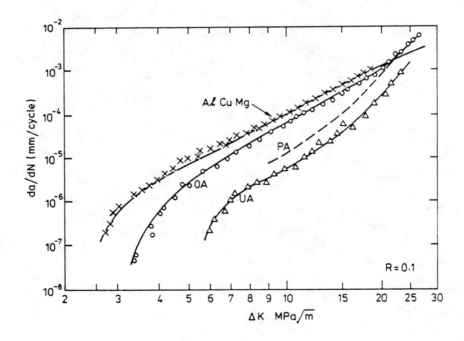

Fig. 24. Fatigue crack growth in AlLiCuMgZr in different aged conditions (OA,PA,UA) compared with AlCuMg alloy (after Harris, Noble and Dinsdale (1984))

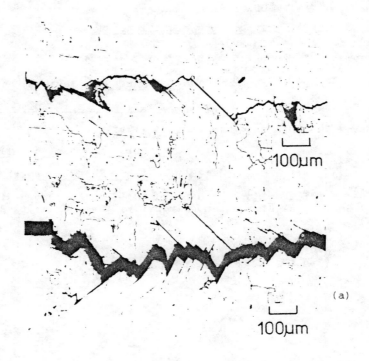

Fig. 25.   Crack path in (a) underaged and (b) overaged
conditions of AlLiCuMgZr alloy (after Harris,
Noble and Dinsdale (1984)

Lankford, J., (1983).   Fat. Eng. Mats. and Strucs. 6, 15-32.

Lindley, T.C., Richards, C.E. and Ritchie, R.O. (1976), Metallurgia and Metal
Forming 43, 268-272.

Lindley, T.C., McIntyre, P. and P.J. Trant (1982) Metals Technology, 9, 135-142.

Lindley, T.C. (1985) Advanced Seminar on Fracture Mechanics Ispra Italy,
p. 167-213, Applied Science Publishers.

Marshall, P.   Private Communication (1985).

Minakawa, K. and McEvily, A.J. (1981).   Scripta Met. 15, 633-642.

Morris, W.L., Buck, O. and Marcus, H.L. (1976).   Met. Trans. A7, 1161-1165.

Morris, W.L., James, M.R. and Buck, O., (1981).   Met. Trans., 12A, 57-64.

Pearson, S. (1975).   Eng. Frac. Mech. 7, 235-247.

Priddle, E.K. (1978).   Scripta Met., 12, 49-56.

Ritchie, R.O. (1979).   Int. Met. Rev. No. 245, 29, 205-230.

74

Schijve, J. (1979). Eng. Fract. Mech., 11, 167-224.

Schijve, J. (1984). Proceeding of Fatigue 84, 3-7 Sept. 1984, Birmingham University, England 751-772, Publishers EMAS.

Suresh, S., (1983). Met. Trans. A, 14A, 2375-2385.

Suresh, S. and Ritchie, R.O. (1984). Int. Met. Reviews, 29, 445-476.

Taylor, D. and Knott, J.F. (1981). Fat. Eng. Mats. and Strucs., 4, 147-155.

Thomason, P.F. Private communication (1985).

Ward-Close, C.M. and Beevers, C.J. (1980). Met. Trans., 11A, 1007-1017.

Weston, T. and Wilson, R.N. (1981). RAE Technical Report MAT 373.

Yan, B.D., Farrington, G.C. and Laird, C. (1984). Proceedings of Fatigue 84, 3 to 7 Sept. 1984 Birmingham University, England 1435-1446, Publishers EMAS.

Fatigue Crack Growth — the Complications

N. A. FLECK

Cambridge University Engineering Department, Cambridge, UK

ABSTRACT

A vast body of experimental data has been accumulated on the constant amplitude crack growth response of structural metals in moist laboratory air. Usually the data is presented as plots of crack growth rate, da/dN, against stress intensity range, ΔK. In order to extrapolate this data to fatigue crack growth in more active or more inert environments, to crack growth under variable amplitude loading, or to crack growth under multi-axial or mixed Mode loading, the mechanisms of crack advance and crack closure should be considered.

This paper briefly reviews the crack closure phenomenon and discusses the dominant causes of accelerated and retarded growth under changes in environment or type of loading. It is argued that simple constant amplitude data is often surprisingly accurate when used to predict crack growth in more complex situations. However, there are some cases where constant amplitude data lead to dangerously non-conservative predictions of fatigue life.

KEYWORDS

Fatigue crack growth; crack closure: environmental effects; variable amplitude loading; multi-axial loading; mixed Mode loading.

INTRODUCTION

The fatigue life of many engineering components is determined by the number of cycles required to grow a single crack to a critical length. Prediction of crack growth rate is therefore important. Although crack propagation rate, da/dN, is strongly dependent on stress intensity range ΔK, it is also a function of many other variables. Material composition and microstructure, the environment, load history and nature of loading (uniaxial or multi-axial) all influence crack growth rate and, thereby, fatigue life.

In the present paper, the following questions are addressed:

1   What physical mechanisms account for the influences of environment, load history and nature of loading on crack growth rate?
2   How can we use simple fatigue crack growth data from constant amplitude tests conducted in "laboratory air" to predict crack growth rates in more complex situations?

It is found that the phenomenon of crack closure plays a key but not the only role in answering these questions.

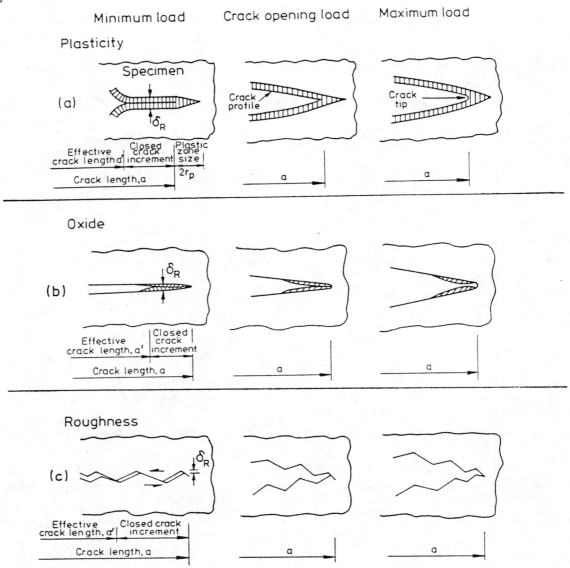

Figure 1: Three forms of crack closure. The residual displacement, δ_R, is the 'excess' material wedged between the crack flanks.

CRACK CLOSURE

In the late 1960's Elber observed the surprising result that fatigue cracks close at above zero load. He investigated thin centre-cracked panels made from 2024-T3 aluminium alloy, and found that the fraction of the load cycle for which the crack is open, U, may be as low as 0.5 for zero-tension loading, Elber (1970, 1971). He was able to account for the increase in constant amplitude growth rate with increasing mean stress by assuming that crack advance is governed by the portion of the stress intensity range for which the crack is open, ΔK_{eff}, where

$$\Delta K_{eff} = U \ \Delta K$$

Elber suggested that residual plastically-stretched material is left on the crack flanks when a fatigue crack advances through the plastic zone at its tip, Fig. 1(a). This extra volume of material between the crack faces leads to premature closure of the crack. Such closure is termed plasticity-induced crack closure. "Extra material" is also found on the crack flanks when severe oxidation of the crack flanks occurs due to fretting processes just behind the crack tip, Paris et al (1972), Stewart (1980), and Suresh et al (1982): this is termed oxide-induced crack closure, Fig. 1(b). Fracture surface roughness can also lead to premature contact of the crack flanks, Halliday and Beevers (1979): this phenomenon is called roughness-induced crack closure, see Fig. 1(c).

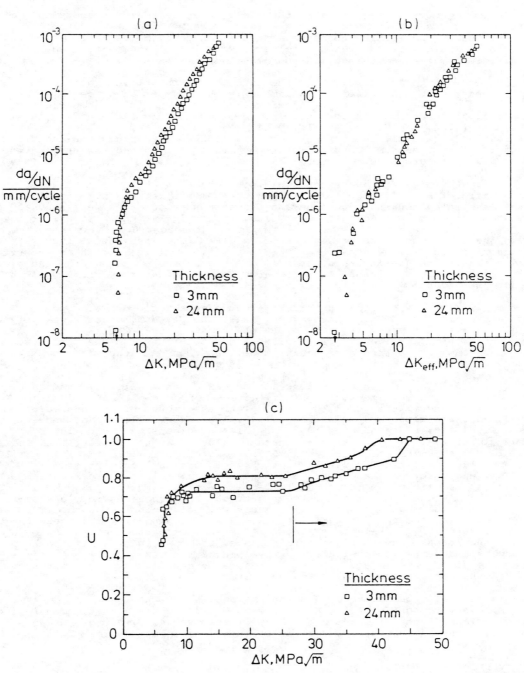

Figure 2: Effect of thickness on crack growth and closure behaviour of BS4360 50B steel. R = 0.05. Compact tension geometry. The arrow in Fig. 2(c) indicates where the forward plastic zone size exceeds the size criterion of w/25, w being the specimen width, 50 mm.

When crack growth rates are near-threshold (that is, da/dN < about 10^{-6} mm/cycle), fatigue cracks tend to grow, using Forsyth's terminology, in a Stage I manner, and mixed Mode I and II displacements are experienced by material in the crack tip region. The associated residual plastic Mode II displacements lead to fracture surface mismatch and thence to roughness-induced crack closure. Cyclic Mode II displacements induce locking of the asperities on the crack faces and thus cause further roughness-induced closure. When the environment is active, the cyclic Mode II displacements also lead to fretting damage in the form of an oxide layer on the fracture surfaces. The thickness of this oxide layer is found to be comparable with the Mode I crack tip opening displacement, and so oxide-induced closure is also significant, Stewart (1980) and Suresh et al (1982). As growth rates approach the threshold rate (about 10^{-7} mm/cycle), the amount of oxide-induced and roughness-induced closure usually outweighs the contribution from plasticity-induced closure and the closure value U often drops to near zero, for example Fig. 2.

When growth rates are above threshold, the crack advances in a Stage II manner and the ratio of Mode II/Mode I near-tip displacements reduces to a negligible level. The associated roughness-induced and oxide-induced closures decrease, and plasticity-induced closure then dominates. Typically, for low strength steels at a load ratio, R (= $\sigma_{min}/\sigma_{max}$), of 0.05, the closure value U is about 0.7 for plane stress conditions and about 0.8 for plane strain conditions, Fig. 2. When the maximum stress intensity factor of the fatigue cycle, K_{max}, is sufficiently large for the plane stress plastic zone size to be greater than about $(1/25)^{th}$ of the specimen width, the elastic field around the crack is insufficiently large to close the crack firmly and the U-value increases to unity, see Fig. 2 and Fleck (1984a).

It is widely accepted that plasticity-induced crack closure can occur under plane stress conditions. Under these conditions, the residual plastic wake of material on the crack flanks comes from the sides of the specimen. There is still debate on whether plasticity-induced crack closure can occur under plane strain conditions: no satisfactory continuum mechanics model has been developed to show how "extra material" can be generated and left on the crack flanks.

Yet, there is experimental evidence to suggest that plasticity-induced crack closure occurs in thick specimens. As ΔK is increased from the threshold value ΔK_{th} to that typical of the mid-range of the Paris plot, the amount of oxide-induced and roughness-induced crack closure decrease and plasticity-induced closure dominates. Mille (1979) and Fleck (1984a) have found that for thick specimens made from low strength steel, the closure value, U, attains a constant value of about 0.8 for ΔK in the range 15-25 MPa$\sqrt{m}$, and a load ratio, R, of 0.05-0.1, Fig. 2. Fleck and Smith (1982) and Fleck (1984a) considered 24 mm thick compact tension specimens made from BS4360 50B steel, and observed the closure response using a variety of compliance gauges. A back face strain gauge, crack mouth gauge and a novel push-rod gauge all indicated that the U-value was 0.8 at the centre of the specimens, where plane strain conditions prevailed. Clip gauges and strain gauges were also mounted on the side faces of the specimens so that they straddled the crack just behind its tip; these gauges indicated a U-value of 0.7. The stress state at the surface of these specimens was closer to plane stress than to plane strain; thus, the plastic zone size ahead of the crack tip and the residual plastic wake behind the crack tip were greatest at the surface of the specimens. It is, therefore, expected that the greatest amount of closure should be shown at the surface.

Tests were also performed on 3 mm thick specimens at the same ΔK and R values. These specimens were sufficiently thin for plane stress conditions to prevail along the whole crack front. Interestingly, the closure value, U, along the whole crack front was 0.7, the same as at the surface of the 24 mm thick specimens.

Crack Closure Models

Several attempts have been made to model plasticity-induced, oxide-induced and roughness-induced crack closure. Oxide-induced closure has been modelled simply by assuming that the oxide debris behind the crack tip behaves as a rigid wedge of constant thickness, placed a small distance behind the crack tip, Suresh et al (1982).

Suresh and Ritchie (1982) have also discussed roughness-induced crack closure in terms of a simple geometric model. They idealise the fatigue fracture surfaces as two intermeshing wavy surfaces which match perfectly at the maximum load of the fatigue cycle. At lower loads the surfaces move normal to each other (Mode I displacements) and also parallel to each other (Mode II displacements), until they come into contact and the crack closes. Although the model ignores any contribution to roughness-induced closure from residual plastic Mode II displacements left in the wake of the crack, it does show successfully that the roughness-induced closure load is high when the fracture surfaces are rough and cyclic Mode II displacements are large near the crack tip.

When crack growth rates are near-threshold (that is, $da/dN <$ about 10^{-6} mm/cycle), fatigue cracks tend to grow, using Forsyth's terminology, in a Stage I manner, and Mixed Mode I and II displacements are experienced by material in the crack tip region. The associated residual plastic Mode II displacements lead to fracture surface mismatch and thence to roughness-induced crack closure. Cyclic Mode II displacements induce locking of the asperities on the crack faces and thus cause further roughness-induced closure. When the environment is active, the cyclic Mode II displacements also lead to fretting damage in the form of an oxide layer on the fracture surfaces. The thickness of this oxide layer is found to be comparable with the Mode I crack tip opening displacement, and so oxide-induced closure is also significant, Stewart (1980) and Suresh et al (1982). As growth rates approach the threshold rate (about 10^{-7} mm/cycle), the amount of oxide-induced and roughness-induced closure usually outweigh the contribution from plasticity-induced closure and the closure value U, often drops to near zero, for example Fig. 2.

When growth rates are above threshold, the crack advances in a Stage II manner and the ratio of Mode II/Mode I near-tip displacements reduces to a negligible level. The associated roughness-induced and oxide-induced closures decrease, and plasticity-induced closure then dominates. Typically, for low strength steels at a load ratio, R (= $\sigma_{min}/\sigma_{max}$), of 0.05, the closure value U is about 0.7 for plane stress conditions and about 0.8 for plane strain conditions, Fig. 2. When the maximum stress intensity factor of the fatigue cycle, K_{max}, is sufficiently large for the plane stress plastic zone size to be greater than about $(1/25)^{th}$ of the specimen width, the elastic field around the crack is insufficiently large to close the crack firmly and the U-value increases to unity, see Fig. 2 and Fleck (1984a).

It is widely accepted that plasticity-induced crack closure can occur under plane stress conditions. Under these conditions, the residual plastic wake of material on the crack flanks comes from the sides of the specimen. There is still debate on whether plasticity-induced crack closure can occur under plane strain conditions: no satisfactory continuum mechanics model has been developed to show how "extra material" can be generated and left on the crack flanks.

Yet, there is experimental evidence to suggest that plasticity-induced crack closure occurs in thick specimens. As ΔK is increased from the threshold value ΔK_{th} to that typical of the mid-range of the Paris plot, the amount of oxide-induced and roughness-induced crack closure decrease and plasticity-induced closure dominates. Mille (1979) and Fleck (1984a) have found that for thick specimens made from low strength steel, the closure value, U, attains a constant value of about 0.8 for ΔK in the range 15-25 MPa$\sqrt{m}$, and a load ratio, R, of 0.05-0.1, Fig. 2. Fleck and Smith (1982) and Fleck (1984a) considered 24 mm thick compact tension specimens made from BS4360 50B steel, and observed the closure response using a variety of compliance gauges. A back face strain gauge, crack mouth gauge and a novel push-rod gauge all indicated that the U-value was 0.8 at the centre of the specimen, where plane strain conditions prevailed. Clip gauges and strain gauges were also mounted on the side faces of the specimens so that they straddled the crack just behind its tip; these gauges indicated a U-value of 0.7. The stress state at the surface of specimen is closer to plane stress than to plane strain; thus, the plastic zone size ahead of the crack tip and residual plastic wake behind the crack tip are greatest at the surface of a thick specimen. It is, therefore, expected that the greatest amount of closure should be shown at the surface of the thick specimen.

Tests were also performed on 3 mm thick specimens at the same ΔK and R values. These specimens were sufficiently thin for plane stress conditions to prevail along the whole crack front. Interestingly, the closure value, U, along the whole crack front was 0.7, the same as at the surface of the thick specimens.

Crack Closure Models

Several attempts have been made to model plasticity-induced, oxide-induced and roughness-induced crack closure. Oxide-induced closure has been modelled simply by assuming that the oxide debris bheind the crack tip behaves as a rigid-wedge of constant thickness, placed a small distance behind the crack tip, Suresh et al (1982).

Suresh and Ritchie (1982) have also discussed roughness-induced crack closure in terms of a simple geometric model. They idealise the fatigue fracture surface as two intermeshing wavy surfaces which match perfectly at the maximum load of the fatigue cycle. At lower loads the surfaces move normal to each other (Mode I displacements) and also parallel to each other (Mode II displacements), until they come into contact and the crack closes. Although the model ignores any contribution to roughness-induced closure from residual plastic Mode II displacements left in the wake of the crack, it does show successfully that the roughness-induced closure load is high when the fracture surfaces are rough and cyclic Mode II displacements are large near the crack tip.

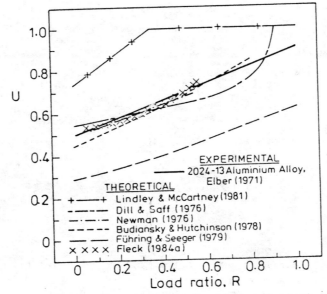

Figure 3: Comparison of plasticity-induced crack closure models with experimental results for 2024-T3 aluminium alloy.

Much more complex analytic and numerical models have been proposed for plasticity-induced crack closure. Newman (1976) and Ogura et al. (1977) have used finite element techniques to model crack closure in 2024-T3 aluminium alloy and a low strength steel, respectively. Newman conducted a plane stress analysis and considered an elastic/perfectly-plastic solid; his results are in good agreement with the measured closure response of 2024-T3 aluminium alloy, see Fig. 3. Ogura et al considered an elastic/work-hardening solid and attempted to model the influence of constraint, yield stress and strain hardening rate on U. They predicted successfully that the plane strain closure value, U, is about 0.8 for a low strength steel at R = 0. Also, they showed that U increases with increasing strain hardening rate, in agreement with the experimental observation that the U-values for steels are higher than for aluminium alloys of lower work hardening rate but similar strength, Fleck (1984a). It is also found experimentally that the closure value, U, increases with increasing yield stress and constraint, Figs. 2 and 4; the finite element model of Ogura et al. was unable to demonstrate this.

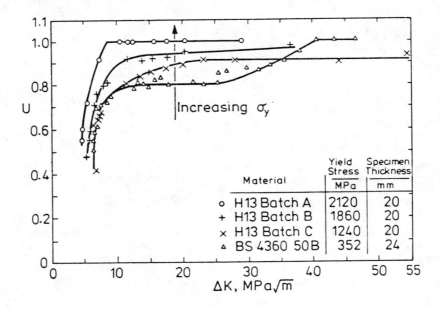

Figure 4: Typical effect of yield stress on closure response of steels, R = 0.05.

Dill and Saff (1976), Budiansky and Hutchinson (1978), Führing and Seeger (1979) and Lindley and McCartney (1981) have modelled plasticity-induced crack closure in an elastic/perfectly-plastic solid by a Dugdale-type idealisation of the crack tip plastic zones. Only plane stress behaviour is considered. Small differences in the assumed boundary conditions and, in particular, in the fraction of the crack tip opening displacement which contributes to the plastic wake, lead to large differences in the predicted closure response, Fig. 3.

Fleck (1984a) has developed a semi-analytic closure model, in which the residual plastic wake along the crack flanks behaves like a giant dislocation centred in the middle of the forward plastic zone. Unlike the other plasticity-induced closure models, this simple semi-analytic model successfully accounts for the influence of yield stress and strain hardening on closure response, as well as the effect of load ratio, R, see Fig. 5. The model can account for the influence of constraint on closure response only if it is calibrated separately for plane stress and for plane strain conditions.

To conclude, the current closure models give some physical insight into the closure phenomenon but require further development.

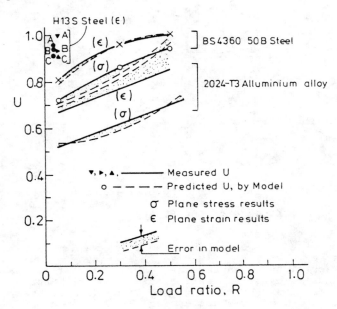

Figure 5: Ability of simple plasticity-induced closure model (Fleck, 1984a) to account for the influence of yield stress, strain hardening rate and load ratio on the closure value U at mid-ΔK values.

Ability of Crack Closure to Rationalise Constant Amplitude Fatigue Crack Growth

It is found experimentally that the threshold stress intensity range, ΔK_{th}, decreases and growth rates increase with increasing mean stress, for example Paris et al. (1972) and Ritchie (1980). This effect is due largely to crack closure: when mean stress is increased, the fatigue crack is open for a larger portion of the fatigue cycle and crack growth rates increase. It is often found that the relation between crack growth rate, da/dN, and effective stress intensity range, ΔK_{eff}, is independent of the mean stress, for example, Fig. 6. A relation of the form

$$\frac{da}{dN} = C(\Delta K_{eff})^m$$

may then be able to correlate the data down to the threshold growth rate of 10^{-7} mm/cycle.

Crack closure is not the only key, however, to an understanding of constant amplitude fatigue. Changes in crack advance mechanism from striation formation to cleavage or microvoid coalescence also result in increased growth rates, Maddox (1975), Lindley et al (1975) and Ritchie and Knott (1973). For example, Ritchie and Knott (1973) have found that the micromechanism of cleavage in embrittled En-30A steel leads to an increase in above threshold growth rate by an order of magnitude when the load ratio, R, is increased from 0 to 0.5. Similarly, Griffiths et al. (1971)

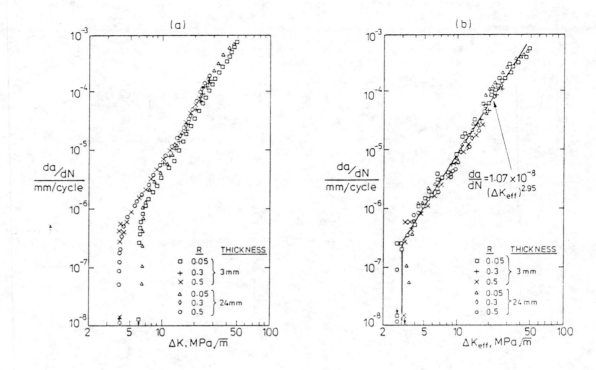

Figure 6: Ability of crack closure to account for the influences of mean stress and thickness on the crack growth response of BS4360 50B steel.

found that microvoid coalescence in a low alloy weld metal leads to an order of magnitude increase in growth rate as the load ratio is increased from 0 to 0.5. It appears that crack closure can account for the mean stress influence on growth rate when crack advance is by striation formation, Fleck (1984a), but is unable to account for mean stress effects when crack growth is by "static" mechanisms of crack advance such as cleavage and microvoid coalescence.

The above discussion has tacitly ignored the influence of environment on the fatigue process. What are the effects of environment on crack growth rate and can crack closure be used to explain them?

EFFECT OF ENVIRONMENT ON CRACK GROWTH RESPONSE

Several investigators have found that the effect of environment on near-threshold growth in steels can be explained in terms of crack closure, Stewart (1980) and Suresh et al. (1982). Crack growth rates are highest and thresholds lowest when tests are performed in a dry, inert atmosphere such as dry argon. In a more active environment such as moist air, the amount of oxide-induced closure increases, leading to higher thresholds and lower growth rates. Crack closure arguments cannot be used, however, to explain the fact that for steels ΔK_{th} in vacuo is higher than in air. Instead, it has been suggested that significant re-welding of the crack growth increment occurs in every cycle, leading to a high threshold in vacuo, Stewart (1980).

There is substantial evidence to suggest that crack closure is <u>unable</u> to explain faster crack growth in active environments, when growth rates are above threshold (that is, above about 10^{-6} mm/cycle). For instance, Ritchie (1980) has noted that hydrogen can lead to accelerated growth in steels by the process of hydrogen embrittlement. When mean stress is increased, the proportion of crack advance by hydrogen-induced intergranular cleavage increases accordingly, and a further increase in growth rate results.

It has also been found that crack closure is unable to account for the influence of environment on above-threshold growth rates in aluminium alloys. Davidson and

Lankford (1983) recently investigated the effect of water vapour on crack advance in 7075-T651 aluminium alloy. They found that the crack tip can sustain much less plasticity in humid air than in dry air: crack closure is not the explanation for the fact that growth rates in humid air are an order of magnitude faster than the corresponding growth rates in dry air. Similarly, Schijve and Arkema (1976) and Ewalds (1980) determined the influence of environment on the crack closure and growth rate responses of 2024-T3 and 7075-T6 aluminium alloys. Growth rates were fastest in salt water and slowest in vacuo, yet the crack opening stress intensity factor, K_{op}, remained unaltered. Bowles (1978) found that the mechanisms of crack advance were different in the two environments, thus providing an explanation for the different growth rates. In inert environments, crack advance is by slip mechanisms, leading to shear cracks and crack branching; in more active environments crack advance is by quasi-brittle cracking followed by crack blunting.

Thus, a variety of physical mechanisms including crack closure account for the influence of environment on fatigue crack growth rate. At present, there are no successful models to predict the effect of material, environment and mean stress on constant amplitude crack growth rate or closure response. Thus, it is still necessary to determine experimentally the relation between da/dN and ΔK for each material under test conditions most closely akin to those experienced in service by engineering components. Usually, the fatigue loading on a structure is not constant in amplitude. Also, the loading is biaxial or multi-axial, rather than uniaxial. Can we predict fatigue lives under these conditions from simple, constant amplitude, uniaxial data?

FATIGUE CRACK GROWTH UNDER VARIABLE AMPLITUDE LOADING

Unfortunately, the crack growth increment per cycle of variable amplitude loading is not equal to the crack growth rate for an equivalent constant amplitude load cycle: retardations and accelerations occur. It is clear that crack advance under variable amplitude loading is dependent upon the previous load history.

There is still much debate over the ability of crack closure to account for load-interaction effects due to variable amplitude loading. The subject has been reviewed by Schijve (1980), Broek (1982), Fleck (1984b, 1985) and Fleck and Smith (1984).

Crack growth retardation is observed when a single peak overload is injected into a constant amplitude load history, for steels, aluminium alloys and titanium alloys. Elber (1971) has suggested that the crack growth retardation is caused by an increase in K_{op} as the crack grows through the overload plastic zone, Fig. 7. Several investigators have confirmed Elber's findings when the pre-overload growth rate is far above threshold, Tanaka et al. (1981), Lankford and Davidson (1981) and Fleck et al. (1983). When the pre-overload growth rate is near-threshold, the overload leads to immediate retardation (or arrest) and to a decreased K_{op}, de Castro and Parks (1982); the cause of this retardation is not known.

The retardation in growth rate following a step decrease in maximum load, and acceleration associated with a step increase in maximum load have also been attributed to crack closure, see Elber (1971), Trebules et al. (1973) and Fig. 7. Immediately after the step change in loading, the closure load gradually changes to the level characteristic of constant amplitude loading at the new load amplitude and a growth rate transient results, Fig. 7. Little experimental data is available to support or refute these suggestions.

Other load-interaction phenomena experienced by steels and aluminium alloys include accelerated growth associated with periodic underloads and retarded growth associated with periodic overloads, de Jonge and Nederveen (1980), Stephens et al. (1977) and Fleck (1985). Fleck (1985) has found that these small accelerations and retardations of less than a factor of two in growth rate are not due to crack closure. Rather, a combination of other mechanisms are responsible, including strain hardening of material ahead of the crack tip due to the underloads, crack tip blunting due to the overloads and a change of the local mean stress ahead of the crack tip induced by the underloads or overloads.

The load-interaction effects associated with a service load history are similar in form to the interaction effects associated with the simpler tests described above, Schijve (1980). For example, severe gust loading on an aircraft structure is similar to single or periodic overloads and leads to crack growth retardation. When high loads occur only occasionally in the load history the crack opening stress intensity, K_{op}, varies throughout the load history and transient retardations and accelerations are observed, see Schijve (1979), Fleck and Smith (1984) and Fig. 7. However, when the load history is repeated over a crack growth increment of much less than the maximum forward plastic zone size generated by the load history, K_{op} remains reasonably constant from one load cycle to the next, Fleck and Smith (1984).

It is found experimentally that crack closure is able to account for crack growth

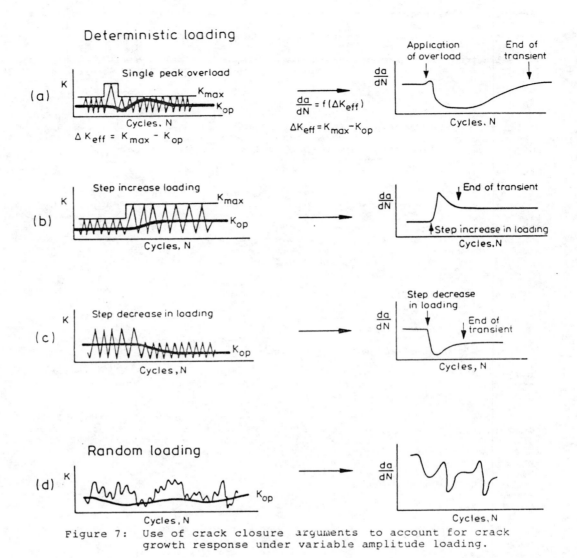

Figure 7: Use of crack closure arguments to account for crack growth response under variable amplitude loading.

rates under service loadings to within a factor of about two, Socie (1977), Kikukawa et al (1981) and Fleck and Smith (1984). Thus, the closure-based fatigue life prediction models of Newman (1981) and de Koning (1981) have the correct physical basis. Such closure models are based upon the reasonable supposition that the constant amplitude relation between crack growth rate, da/dN, and ΔK_{eff} is preserved for variable amplitude loading: they assume that any influence of material or environment is encapsulated within the constant amplitude da/dN - ΔK_{eff} relation.

It is concluded that crack closure is the main cause of load interaction effects under variable amplitude loading. Constant amplitude da/dN - ΔK_{eff} data may be used to predict fatigue life provided the crack closure response, K_{op}, is known experimentally or theoretically for the service load history.

MULTI-AXIAL AND MIXED MODE LOADINGS

In the following discussion multi-axial loading is defined to be an applied load system where the resolved stresses on the cracking plane have no Mode II or Mode III components. Mixed Mode loading is such that the resolved stresses on the cracking plane give rise to Mode II or Mode III displacements at the crack tip.

Consider first multi-axial loadings. Several investigators have measured crack growth rates of through-cracks in plates subjected to remote biaxial loading, Fig. 8. See Smith and Pascoe (1983) for a recent review of these investigations.

There seems to be no universal agreement over the influence of the stress range, ΔS_x

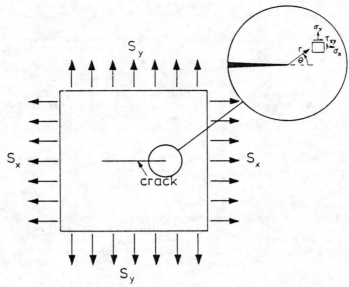

Figure 8: Biaxial loading of a centre-cracked plate.

applied parallel to the cracking plane on the fatigue crack growth rate. For example, Joshi and Shewchuk (1970) found that a cyclic, fully tensile stress range, ΔS_x, in phase with the stress range, ΔS_y, Fig. 8, leads to an increase in crack growth rate in 2024-T351 aluminium alloy. Brown and Miller (1980), however, found that a tensile ΔS_x in phase with ΔS_y led to a decrease in crack growth rate, compared with the simple tension case, in 316L stainless steel. Also, Liu et al (1978) found that ΔS_x had no influence on crack growth rate, whether it was fully tensile or fully compressive; they considered 2024-T351 and 7075-T7351 aluminium alloys.

It is easily seen from theoretical considerations that the stress ΔS_x may have a small influence on the crack growth rate. The asymptotic Mode I elastic solution for the stress field near the tip of a crack in a biaxial stress field is given by (Williams, 1957; Eftis et al, 1977),

$$\Delta \sigma_y = \frac{\Delta K_I}{\sqrt{2\pi r}} \; \cos \frac{\theta}{2} \left(1 + \sin \frac{\theta}{2} \sin \frac{3\theta}{2} \right) + O(r^{\frac{1}{2}})$$

$$\Delta \sigma_x = \frac{\Delta K_I}{\sqrt{2\pi r}} \; \cos \frac{\theta}{2} \left(1 - \sin \frac{\theta}{2} \sin \frac{3\theta}{2} \right) + \Delta T + O(r^{\frac{1}{2}})$$

$$\Delta \tau_{xy} = \frac{\Delta K_I}{\sqrt{2\pi r}} \; \sin \frac{\theta}{2} \cos \frac{\theta}{2} \cos \frac{3\theta}{2} + O(r^{\frac{1}{2}})$$

where σ_y, σ_x, τ_{xy}, r and θ are defined in Fig. 8.

The stress intensity range, ΔK_I, is given by

$$\Delta K_I = \Delta S_y \; \sqrt{a} \; f(a, \text{geometry})$$

where ΔS_y is the remote stress range in the y-direction, a is the crack length and the calibration function, f, is well-known for many specimen geometries.

The non-singular term, ΔT, in the expression for $\Delta \sigma_x$ is given by,

$$\Delta T = \Delta S_x \; g_1(a, \text{geometry}) + \Delta S_y \; g_2(a, \text{geometry})$$

For the case of a centre-crack in an infinite sheet, g_1 equals 1 and g_2 equals -1, from Eftis et al (1977).

It is clear from the above equations that the stress range ΔS_x has no influence on ΔK_I, but does influence ΔT and thereby $\Delta \sigma_x$ and the effective stress range in the vicinity of the crack tip. By this argument Eftis et al (1977) show that a fully tensile ΔS_x leads to smaller effective stress ranges and to smaller plastic zone sizes than for the case of uniaxial tension. Similar conclusions have been drawn by

Miller and Kfouri (1974), using an elastic-plastic finite element scheme.

Thus, any small decrease in crack growth rate in the presence of a fully tensile ΔS_x may be due to a decrease in the crack tip plastic zone size. Theoretical studies have also shown that a fully tensile ΔS_x leads to smaller crack tip openings (Adams, 1973 and Miller and Kfouri, 1974), and to a higher crack opening load (Ogura et al, 1974). Further work is required to determine the dominant cause of any accelerations or retardations associated with biaxial loading.

From the above discussion of crack tip stresses it appears that the relevant parameters to characterise fatigue crack growth in biaxial stress fields are ΔK_I and ΔT. Unfortunately, most investigators assume that ΔK_I and the biaxiality ratio S_x/S_y are the relevant parameters. The many discrepancies reported in the literature on the effects of biaxial loading on crack growth rate may arise from this basic misunderstanding.

Generally, it is found that the degree of biaxiality has only a small influence on crack growth rates: simple uniaxial data often suffice for accurate estimates of fatigue life.

Now consider mixed Mode loading. Tschegg et al (1983) have applied pure Mode III loading to cylindrically-notched bar specimens made from AISI 4340 steel. They found that the crack growth rate at constant ΔK_{III} decreased from a rate somewhat higher than the equivalent Mode I growth rate to crack arrest as the circumferential cracks grew towards the centre of the specimens. The decreasing growth rate with increasing crack length was attributed to severe rubbing and interlocking between the fracture surfaces. Smith and Smith (1984) have examined the crack growth response of BS4360 50B structural steel to pure Mode II loading and found that the crack surface interactions are even more severe than under pure Mode III. When the Mode II stress intensity range exceeded about 19 MPa$\sqrt{m}$ at a load ratio, R, (= K_{IImin}/K_{IImax}) of 0.05, full cyclic slip occurred along the crack flanks to the crack tip and a Mode I branch crack was formed at the crack tip. Cyclic slip did not reach the crack tip at lower ΔK_{II} levels, and no growth occurred. The Mode I threshold stress intensity range for this material is 6.0 MPa$\sqrt{m}$ for a load ratio, K_{Imin}/K_{Imax}, of 0.05.

Plainly, crack closure in the form of fracture surface rubbing leads to increased fatigue lives under Mode II or Mode III loading, compared with Mode I loading. Mode I crack growth data is conservative in most instances.

CONCLUSIONS

A correct physical picture of the fatigue cracking process in metallic systems must include a knowledge of the mechanisms by which cracks advance and a knowledge of the crack closure phenomenon. The crack advance mechanism and the amount of crack closure may change with environment or with the type of loading, leading to a corresponding change in the crack growth rate.

Constant amplitude crack growth data taken in moist air is often conservative when used to predict crack growth under variable amplitude loading, multi-axial loading, or mixed Mode loading. However, this constant amplitude data is not conservative when used to predict the following:

1   Near threshold growth rates in steels when the environment is inert, for example, dry argon.

2   Above threshold growth rates in steels and aluminium alloys, when the environment is highly active, such as gaseous hydrogen for steels and salt water for aluminium alloys.

3   Crack growth rates in steels or aluminium alloys when the loading consists of periodic underloads.

These conditions promote accelerations in crack propagation rate due to several mechanisms, including crack closure, changes in the micro-mechanism of crack advance and strain hardening of material at the crack tip.

REFERENCES

Adams, N J I (1973). Some comments on the effect of biaxial stress on fatigue crack growth and fracture, Engng. Fracture Mech., 5, 983-991
Bowles, C Q (1978). PhD Thesis, Delft University
Broek, D (1982). Elementary Engineering Fracture Mechanics, 2nd Edition, Martinus Nijhoff Publishers, The Hague
Brown, M W and K J Miller (1980). Development of a biaxial facility for testing high temperature materials, NRC Report, Cambridge, England

Budiansky, B and J W Hutchinson (1978). Analysis of closure in fatigue crack growth, J. Appl. Mech., 45, 267-276

Davidson, D L and J Lankford (1983). The effect of water vapor on fatigue crack tip mechanics in 7075-T651 aluminium alloy, Fatigue of Engng. Mats. and Structures, 6, 3, 241-256

de Castro, J T P and D M Parks (1982). Decrease in closure and delay of fatigue crack growth in plane strain, Scripta Met., 16, 1443-1446

de Jonge, J B and A Nederveen (1980). Effect of load variables on fatigue crack initiation and propagataion, ASTM STP714, 170

de Koning, A U (1981). A simple crack closure model for prediction of fatigue crack growth rates under variable-amplitude loading, Fracture Mechanics: Thirteenth Conference, ASTM STP743, 63-85

Dill, H D and C R Saff (1976). Spectrum crack growth prediction method based on crack surface displacement and contact analyses, Fatigue Crack Growth Under Spectrum Loads, ASTM STP595, 306-319

Eftis, J, N Subramonian and H Leibowitz (1977). Crack border stress and displacement equations revisited, Engng. Fracture Mech., 9, 189-210

Elber, W (1970). Fatigue crack closure under cyclic tension, Engng. Fracture Mech., 2, 37-45

Elber, W (1971). The significance of fatigue crack closure, ASTM STP486, 230-242

Ewalds, H L (1980). The effect of environment on fatigue crack closure in aluminium alloys, Engng. Fracture Mech., 13, 1001-1007

Fleck, N A (1984a). An investigation of fatigue crack closure, Report CUED/C/MATS/TR104 (PhD Thesis), Cambridge University, UK, May 1984

Fleck, N A (1984b). Influence of stress state on crack growth retardation, presented at ASTM Workshop on Fundamental Questions and Critical Experiments on Fatigue, Dallas, Texas, October 18-20, 1984

Fleck, N A (1985). Fatigue crack growth due to periodic underloads and periodic overloads, to appear in Acta Metallurgica

Fleck, N A, I F C Smith and R A Smith (1983). Closure behaviour of surface cracks, Fatigue of Engng. Mats. and Structures, 6 (3), 225-239

Fleck, N A and R A Smith (1982). Crack closure - is it just a surface phenomenon? Int. J. Fatigue, July 1982, 157-160. Correction in Int. J. Fatigue, Oct. 1982, 243

Fleck, N A and R A Smith (1984). Fatigue life prediction of a structural steel under service loading, Int. J. Fatigue, 6 (4), 203-210

Führing, H and T Seeger (1979). Dugdale crack closure analysis of fatigue cracks under constant amplitude loading, Engng. Fracture Mech., 11, 99-122

Griffiths, J R, I L Mogford and C E Richards (1971). Effect of mean stress on fatigue-crack propagation in a ferritic weld metal, Metal Science, 5, 150-154

Halliday, M D and C J Beevers (1979). Non-closure of cracks and fatigue crack growth in β-heat treated Ti-6Al-4V, Int. J. Fracture, 15, R27-R30

Joshi, S R and J Shewchuk (1970). Fatigue-crack propagation in a biaxial stress field, Expl. Mech., 10, 529-533

Kikukawa, M, M Jono and Y Kondō (1981). An estimation of fatigue crack propagation rate under varying load conditions of low stress intensity level, Advances in Fracture Research, Proc. 5th Int. Conf. on Fracture, Cannes, France, 29 March - 3 April 1981, ed. D Francois (Pergamon Press, Oxford), 4, 1799-1806

Lankford, J and D L Davidson (1981). The effect of overloads upon fatigue crack tip opening displacement and crack tip opening/closing loads in aluminium alloys, Advances in Fracture Research, Proc. of 5th Int. Conf. on Fracture, Cannes, 899-906

Lindley, T C and L N McCartney (1981). Mechanics and mechanisms of fatigue crack growth, Developments in Fracture Mechanics - Vol. 2, Ed. G G Chell, Appl. Science Publ., London, 247-322

Lindley, T C, C E Richards and R O Ritchie (1975). The mechanics and mechanisms of fatigue crack growth in metals, Central Electricity Research Laboratories, Leatherhead, Surrey, UK, Lab. Note RD/L/N 208/75

Liu, A F, J E Allison, D F Dittmer and J R Yamane (1978). Effect of biaxial stresses on crack growth, 11th Nat. Symp. on Fract. Mech., Blacksburg, VA, June 1978

Maddox, S (1975). The effect of mean stress on fatigue crack propagation: a literature review, Int. J. Fracture, 11 (3), 389-408

Mille, P (1979). Phenomene de fermeture a la pointe de fissure de fatigue dans le cas des aciers, PhD Thesis, Universite de Technologie de Compiegne

Miller, K J and A P Kfouri (1974). An elastic-plastic finite element analysis of crack tip fields under biaxial loading conditions, Int. J. Fracture, 10, 393-404

Newman, Jr., J C (1976). A finite-element analysis of fatigue crack closure, Mechanics of Crack Growth, ASTM STP590, 281-301

Newman, Jr., J C (1981). A crack-closure model for predicting fatigue crack growth under aircraft spectrum loading, Methods and Models for Predicting Fatigue Crack Growth Under Random Loading, ASTM STP748, 53-84

Ogura, K, K Ohji and K Honda (1977). Influence of mechanical factors on fatigue crack clsoure, Fracture 1977, ICF4, Waterloo, Canada, 19-24 June, 2, 1035-1047

Ogura, K, K Ohji and Y Ohkubo (1974). Fatigue crack growth under biaxial loading, Int. J. Fracture, 10, R609-610

Paris, P C, R J Bucci, E T Wessel, W G Clarke and T R Mager (1972). Extensive study of low fatigue crack growth in A533 and A508 steels, Stress Analysis and Growth of Cracks, ASTM STP513, 141-176

Ritchie, R O (1980). Application of fracture mechanics to fatigue, corrosion-fatigue and hydrogen-embrittlement, Analytical and Experimental Fracture Mechanics, Proc. Int. Conf., Rome, Ed. G C Sih, Sijthoff and Noordhoff

Ritchie, R O and J F Knott (1973). Mechanisms of fatigue crack growth in low alloy steel, Acta Met., 21, 639-648

Schijve, J (1979). Four lectures on fatigue crack growth, Engng. Fracture Mech., 11, 167-221

Schijve, J (1980). Prediction methods for fatigue crack growth in aircraft material, Fracture Mechanics: Twelfth Conference, ASTM STP700, 3-34

Schijve, J and W J Arkema (1976). Crack closure and the environmental effect on fatigue crack growth, Delft Univ. Rep. VTH-217

Smith, E W and K J Pascoe (1983). The behaviour of fatigue cracks subject to applied biaxial stress: a review of experimental evidence, Fatigue of Engng. Mats. and Structures, 6 (3), 201-224

Smith, M C and R A Smith (1984). Towards an understanding of Mode II fatigue crack growth, presented at ASTM Symposium on Fundamental Questions and Critical Experiments on Fatigue, Dallas, Texas, 22-23 October, 1984

Socie, D F (1977). Prediction of fatigue crack growth in notched members under variable amplitude loading histories, Engng. Fracture Mech., 9, 849-865

Stephens, R I, E C Sheets and G O Njus (1977). Fatigue crack growth and life predictions in Man-Ten steel subjected to single and intermittent tensile overloads, Cyclic Stress-Strain and Plastic Deformation Aspects of Fatigue Crack Growth, ASTM STP637, 176-191

Stewart, A T (1980). The influence of environment and stress ratio on fatigue crack growth at near threshold stress intensities in low-alloy steels, Engng. Fracture Mech., 13, 463-478

Suresh, S, D M Parks and R O Ritchie (1982). Crack tip oxide formation and its influence on fatigue thresholds, Proc. 1st Int. Symp. on Fatigue Thresholds, Stockholm, Eds. J Bäcklund, A F Blom and C J Beevers, EMAS Publ. Ltd., Warley, UK 1, 391-408

Suresh, S and R O Ritchie (1982). A geometric model for fatigue crack closure induced by fracture surface roughness, Met. Trans A, 13A, 1627

Tanaka, K, S Matsuoka, V Schmidt and M Kuna (1981). Influence of specimen geometry on delayed retardation phenomena of fatigue crack growth in HT80 steel and A5083 aluminium alloy, Advances in Fracture Research, Proc. of 5th Int. Conf. on Fracture, Cannes, 4, 1789-1798

Trebules, Jr., V W, R Roberts and R W Hertzberg (1973). Effect of multiple overloads on fatigue crack propagation in 2024-T3 aluminium alloy, Progress in Flaw Growth and Fracture Toughness Testing, ASTM STP536, 115-146

Tschegg, E K, R O Ritchie and F A McClintock (1983). On the influence of rubbing fracture surfaces on fatigue crack propagation in Mode III, Int. J. Fatigue, 5 (1), 29-35

Williams, M L (1957). On the stress distribution at the tip of a stationary crack, Trans. ASME, J. Appl. Mech., 24, 109-114

A Consideration of the Significant Factors Controlling Fatigue Thresholds

C. J. BEEVERS* and R. L. CARLSON**

*Department of Metallurgy and Materials, University of Birmingham, Birmingham, UK
**School of Aerospace Engineering, Georgia Institute of Technology, Atlanta, Georgia, USA

ABSTRACT

The development of fatigue damage can be one of the factors limiting the useful life of a component or structure. In many instances the fatigue damage is present in the form of small defects which originate from microstructural or fabrication flaws.

The development of these small defects involves, in many instances, stress intensities near the threshold regime and fatigue crack growth rates in the range 10^{-5} to 10^{-10} mm/cycle. As most of the "lifetime" is spent in this low growth rate regime it is necessary to appreciate the influence of variables such as microstructure, stress state and environment on the fatigue response. The interactive role of these variables on both fatigue crack growth rate and ΔK_{th} will be examined. In particular the factors which dominate the fatigue threshold stress intensity level will be identified and considered in detail.

INTRODUCTION

The resistance to growth of defects in the early stages of fatigue life can be related to the fcg (fatigue crack growth) response in the low stress intensity range regime. In particular, cracks which are matrix initiated or arise from other microstructural features, e.g., pores, inclusions, intermetallic particles, can undergo crack growth where the fcg is low and the stress intensity range is in the region of the fatigue threshold. The fatigue threshold ΔK_{th} is considered to be a lower bound value for ΔK below which fcg does not occur. In practical terms ΔK_{th} is defined as the stress intensity range at which fcg rate is 10^{-8} mm/cycle.

The growth of fatigue cracks in the low stress intensity range can be described by a relationship of the form:

$$\frac{da}{dN} = D(\Delta K - \Delta K_{th})^n \tag{1}$$

The form of the fcg versus ΔK curve and the magnitude of the fcg rate for a particular ΔK is dominated by the value of ΔK_{th} in this regime. ΔK_{th} can be considered to have two components, ΔK_{th}^i the intrinsic resistance to fatigue crack growth extension and ΔK_{th}^c the closure contribution

where

$$\Delta K_{th} \text{ (measured)} = \Delta K_{th}^i + \Delta K_{th}^c \tag{2}$$

$$\Delta K^c_{th} = K_{op} - K_{min} \qquad\qquad (3)$$

$$\Delta K^i_{th} = K_{max} - K_{op} \qquad\qquad (4)$$

where K_{op} is the stress intensity necessary to sustain fully opened crack faces with no load transfer behind the crack tip.

It is pertinent at this stage to comment upon the mode of fcg that can be expected at low ΔK's. In many metals and alloys the crack growth process involves separation on crystallographic planes (Ward-Close, 1980; Beevers, 1977) at fcg rates less than $\sim 10^{-5}$ mm/cycle. The separation of these planes in random oriented polycrystalline material leads to a non-uniform trajectory of the crack path through the material. Studies on α-titanium (Walker, 1980) showed that a combination of out of plane trajectories and mode II displacements can lead to a wedging open of the crack faces by the creation of asperities on the fracture faces.

In quenched and tempered steels the fcg process involves both intergranular and transgranular separation (Cooke, 1975). In those materials where the fcg mode leads to a relatively smooth fracture surface (at the micron level) oxide build up near the crack tip has been reported by Suresh (1981) and considered responsible for the wedging open of the fatigue fracture surfaces. The oxide layer builds up to a thickness of the order of 0.1 to 0.2 μm in the threshold region. Thus, the occurrence of crystallographic plane separation or oxide build up on the fracture surface or indeed a combination of both can lead to the creation of asperities on the fatigue fracture surfaces of most materials.

THE EXTRINSIC CLOSURE CONTRIBUTION TO ΔK_{th}

The presence of irregularities or asperities on the fatigue fracture surface can lead to load transfer at points of contact on the fracture face and to modification of the stress intensity range experienced by material ahead of the crack tip.

The Model

Under dominantly plane strain loading conditions at low stress intensities and mean stress levels, the presence of asperities on the fatigue fracture faces can lead to a wedging open of the fatigue crack in the crack tip region. The non-closure of the fatigue crack faces occurs as a direct result of the presence of the asperities.

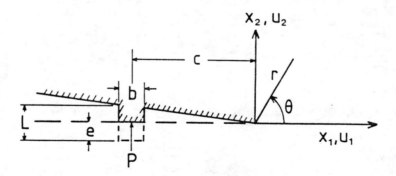

Fig. 1. Crack tip geometry and loading of upper crack face (Beevers, to be published).

Examination of fatigue fracture surfaces from the threshold region reveals the presence of a non-uniform distribution of microstructural asperities. However, if the specimen thickness is very much greater than the spacing of the asperities it is possible to envisage a line of asperities behind the crack tip acting as a load bearing surface. In the case of oxide layer build-up this can be envisaged as creating a relatively uniform asperity across the specimen width. There is, therefore, the possibility of representing the asperities through thickness by a precompressed spring which makes line contact through the thickness of the sample. The line contact concept permits the problem to be treated as a two-dimensional one. Thus, the model to be described treats the case for Mode I opening with a single asperity behind the crack tip under dominantly plane strain conditions (Beevers, to be published).

The essence of the crack tip region is illustrated in Fig. 1. The upper half of the crack tip region is represented; an asperity of width b is positioned at a distance C from the crack tip. L represents the magnitude of the interference between the faces produced by the presence of the asperity. The local crack face force P results in a compression of the asperity of magnitude e.

From consideration of the global and local forces on the crack the following relationships have been developed (Beevers, to be published):

$$K_{cl} = \left(\frac{2}{\pi c}\right)^{\frac{1}{2}} \left[\frac{1}{Eb} + \frac{2}{\pi LG}(1-\nu)\right]^{-1} \tag{5}$$

$$K_{op} = \frac{LG}{2(1-\nu)}\left(\frac{2\pi}{c}\right)^{\frac{1}{2}} \tag{6}$$

where K_{op} is the global stress intensity above which there is no contact between the opposing crack faces and K_{cl} is the local stress intensity which results from the interference with the closing of the faces by the presence of the asperity.

Examination of Equations 5 and 6 reveals that both K_{cl} and K_{op} increase with asperity height and decrease with the distance from the crack tip. With increasing asperity height the crack tip stress intensity will increase and some yielding of the material ahead of the crack tip could be expected. This can be thought of as a form of relaxation of the local stress intensity and reducing the rate of increase of K_{cl} with asperity height. This feature was incorporated into the model by assuming that the local crack tip plasticity would increase the effective value of C. The relationships for K_{op} and K_{cl} now have the form:

$$K_{cl} = \left(\frac{2}{\pi(c+R_p)}\right)^{\frac{1}{2}} \left[\frac{1}{Eb} + \frac{2(1-\nu)}{\pi GL}\right]^{-1} \tag{7}$$

$$K_{op} = \frac{LG}{(1-\nu)}\left(\frac{2\pi}{c+R_p}\right)^{\frac{1}{2}} \tag{8}$$

where $R_p = \frac{1}{6\pi}\left(\frac{K_{max}}{\sigma_y}\right)^2$.

From tests carried out on N901 nickel alloy specimens, K_{cl} was obtained from both load-COD curves and from replica measurements of the crack tip dimensions C, L and b. The two measurements of K_{cl} for this particular case were close to 3.5 MPam$^{\frac{1}{2}}$ and within 15% of each other.

The model also has another feature which is of interest and this can be described by reference to the variation in the stress intensity factor during loading. Analyses of this variation with external load for a compact tensile specimen produce diagrams of the type shown in Fig. 2. For no asperity interference the loading curve for K_1 proceeds along the line $P_o - P_1 - P_2$. If, however, asperity contact prevents complete closure, the model predicts that K_1 for zero external load has a non-zero closure value given by the point P_3. Then, with loading, K_1 values move along the line from P_3 to P_1, which corresponds to the opening load. Further loading then proceeds along the line from P_1 to P_2.

Along the line from P_3 to P_1, the asperity undergoes a decreasing compression to a value of zero at P_1. If the opposing asperities become welded, however, the model then indicates that the load path beyond P_1 proceeds to P_4, and the asperity loading undergoes increasing tension (Carlson, to be published).

The result of these alternative behaviours on the effective stress intensity factor range can be shown by reference to the dashed lines of Fig. 2. Thus, for an external load varying between Q_{min} and Q_{max} with no closure interference, the values of K_1 range from A to B.

For loading with non-closure, K_1 varies from point C to B. That is, the effective ΔK_1 has been reduced by an amount given by the distance between points A and C.

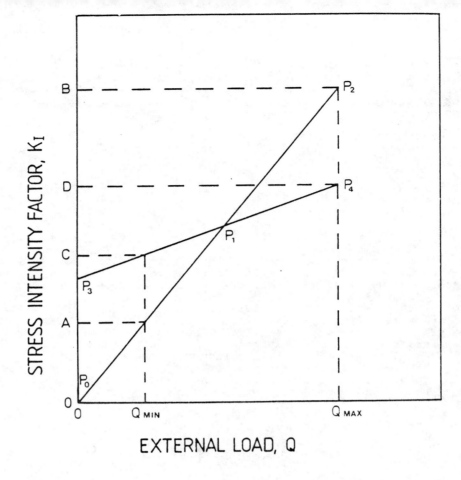

Fig. 2. Stress intensity factor versus external load.

If closure is not complete, and, in addition, asperity welding occurs, K_1 varies between points C and D. This represents a further decrease in the effective ΔK_1.

In the actual fatigue crack problem welding would occur at discrete asperity contacts, rather than on a line contact. Further, some of the weld contacts could be expected to break. If, however, at least some weld contacts remain unbroken, there would be some reduction in the effective ΔK_1 along the crack front.

The mechanism of fracture surface welding has been used as an explanation for experimental results obtained recently by Kendall and Knott (1984). They found that crack growth rates for a low carbon steel were smaller in vacuum than in air. They attribute the difference to greater reversibility of a slip mechanism which leads to rebonding or welding of slip surfaces in a vacuum. In air, contamination of the slipped surfaces inhibits slip reversibility and results in an increment of crack growth.

Another contributing mechanism to this observed difference in crack growth rate can be based on the asperity welding mechanism which leads to a reduction in the effective ΔK_1 as depicted in Figure 2.

Factors Influencing K_{th}^c

Examination of Equation 7 reveals the following features will lead to high values for K_{cl}: large asperities located close to the crack tip, high values for the elastic moduli of the matrix and asperity and high yield stresses for the material ahead of the crack tip.

The projected K_{cl} levels for a range of engineering materials is presented in Figure 3 and Table 1. The curves show that for the parameters chosen for b and C the closure stress intensity K_{cl} approaches an upper bound value dependent upon the matrix material and the asperity. The potential benefits from closure would appear to be higher for nickel and steel compared with aluminium and titanium. The presence of a highly rigid asperity, e.g., Al₂O₃ or TiO₂ has significant influence on K_{cl} and raises the upper bound values into the 7-9 MPam$^{\frac{1}{2}}$ regime.

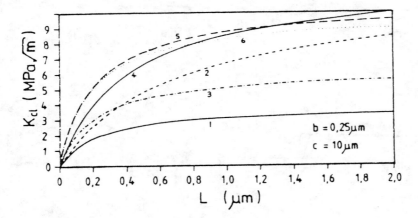

Fig. 3. The projected values of K_{cl} for a range of engineering materials (Table 1 for key information).

TABLE 1 Values used to compute the curves in Figure 3, E and G are in GPa, σ_y in MPa

	Matrix	Asperity	ν	G	E	σ_y
1	Al	Al	.34	26	70	500
2	Al	Al₂O₃	.34	26	370	500
3	Ti	Ti	.36	45	110	850
4	Ti	TiO₂	.36	45	270	850
5	Ni alloy	Ni alloy	.375	76	200	900
6	Steel	Steel	.29	82	210	1000

Thus it would appear that there is the possibility in most engineering alloys to achieve a substantial resistance to fatigue crack growth through control of the character and distribution of the asperities on the fracture faces in the crack tip region.

A Mixed Mode Model

On a microscopic scale, impinging fracture surfaces have been found to have contact on facets inclined to the macroplane of the crack. Suresh (1983 and 1984) developed a crack model which includes features observed in actual cracks when they are magnified. He has used a model of the type shown in Fig. 4(a) to determine the effects of crack deflection away from the Mode I growth plane on crack growth resistance. Recently, Carlson and Beevers (to be published) developed a model which examines conditions developed when the crack tip is between points p and q of Fig. 4(a). Since it has been observed that mode II displacements can occur at the crack tip during loading, a misfit of the inclined surfaces can be introduced. In Fig. 4(b) a shear displacement causing the inclined surfaces to separate is shown. Here, unloading does not produce interference, so contact face forces are not introduced. The mode II displacements can also produce an interference of the type shown in Fig. 4(c). Here, the amount of displacement overlap shown is s. Since the surfaces cannot penetrate one another, an accommodation, which results in forces across the inclined planes, is developed. These forces produce what can be described as a local contribution to the stress state at the crack tip. Superimposed on this state is a contribution due to externally applied loads. In addition to the interference, s, the model includes parameters which give the distance of the jog to the crack tip, C, and the angle of inclination of the jog, φ. Because motion between the contacting faces occurs, friction also needs to be considered. Also, it should be observed that the direction of the friction force during cyclic loading depends on the relative motion of the contacting faces. This means that the friction force undergoes a reversal in direction during each cycle.

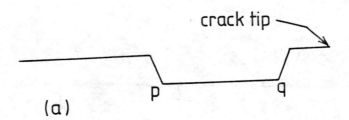

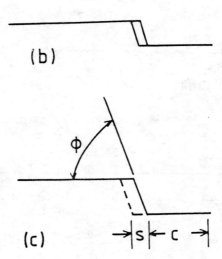

Fig. 4. Nonplanar fracture.

The forces acting on the inclined surfaces can be decomposed into normal and tangential force components, and the loading on the crack faces can be then depicted as shown in Fig. 5. Two dimensional solutions for the locally developed stress intensity factors are available. For the plane strain case the force P_2 produces a mode I stress intensity factor of

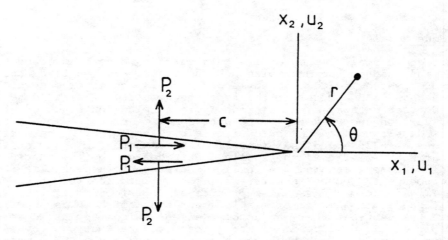

Fig. 5. Crack face loading.

$$K_I(\text{local}) = \left(\frac{2}{\pi c}\right)^{\frac{1}{2}} \frac{P_2}{B},$$

$$(9)$$

where B is the thickness of the cracked body. The mode II stress intensity factor is

$$K_{II}(\text{local}) = \left(\frac{2}{\pi c}\right)^{\frac{1}{2}} \frac{P_1}{B} \ . \tag{10}$$

If the stress intensity factors due to external loading are K_I (global) and K_{II} (global), the total stress intensity factors are, respectively

$$K_I = K_I(\text{local}) + K_I(\text{global}) \ , \tag{11}$$

and

$$K_{II} = K_{II}(\text{local}) + K_{II}(\text{global}) \ . \tag{12}$$

These equations have been used as a basis for the development of a mixed mode closure model (Carlson and Beevers, to be published). By the introduction of appropriate kinematic relations which describe the relative motion of the inclined surfaces, and the use of relations for the displacements of the contact points in terms of local and global values of the stress intensity factors, solutions which give a description of the variation of K_I and K_{II} with varying external loading have been developed (Carlson and Beevers, to be published). Data from fatigue crack propagation experiments on a titanium alloy have been used as a basis for examining this behaviour. For the mechanical properties $G = 41.4$ GN/m^2 and $\nu = 0.34$. A compact tension specimen with W = 50 mm, B = 81 mm, and a = 22.5 mm, was tested. The asperity parameters were taken to be c = 20 μm, $\phi = 50^{\circ}$ and s = 0.839 μm. A friction coefficient of 0.7 was used. The external load used in the computations varied periodically as $Q = Q_m + Q_a \cos(\psi + \pi)$, where ψ varied from 0 to 2 π. The mean value of $Q_m = 5.5$ kN and the alternating value of $Q_a = 4.5$ kN.

The variations in the stress intensity factors during a load cycle are shown in Fig. 6. An examination reveals that the ratio K_I/K_{II} is not constant over the load cycle. The stress state developed near the crack tip is, therefore, non-proportional with reference to loading. This implies, for example, that the orientation of the maximum shear stress at a point would change during a cycle.

The results presented suggest that contact pressures between asperity faces can lead to mixed mode effects which are not insignificant. Non-proportional loading states are produced, and the development of mode two loading creates conditions which promote crack branching.

The closure contact in the two dimensional model is represented as a line load. In the actual closure problem there is a distribution of discrete contacts close to the crack tip. Solutions for normal and tangential point loads have been developed (Kassir, 1973; Meade, 1984), so in principle, it should be possible to model discrete contacts. Although this would be a very difficult task, it is possible to anticipate at least some features of such models. Sliding contact between inclined surfaces could still be expected. In general, however, local K_{III} contributions, as well as those from K_I and K_{II}, could be developed during the closure phase of a loading cycle.

There might be an inclination to suspect that the effects of the local K_{II} and K_{III} stress intensity factors might cancel one another along a crack tip. Since stress singularities are involved, however, exact cancellation could only occur if asperity features satisfy strict conditions that are unlikely to occur.

From these considerations, it can be concluded that the local stress intensity factors can be expected to vary along the crack front during closure. Thus, though the global stress intensity factor may dictate the plane of macro-fracture plane advance, conditions for branching on a micro-scale are developed. If these effects for fatigue crack growth with closure are combined with the crack deflection considerations of Suresh (1983, 1984) it would follow that the crack arrest tendencies associated with threshold behaviour could be promoted.

The Intrinsic Resistance of Materials to FCG and its Contribution to ΔK_{th}

The threshold concept relates to a condition where the crack arrests or the growth rate is so small as not to be detected. A simple but probably effective way of viewing the threshold condition is that it is related to a lower bound value of the crack tip opening displacement (CTOD) which can just sustain fcg:

$$\Delta(\text{CTOD})_c = \frac{(\Delta K_{th}^i)^2}{4\sigma_y' E} \tag{13}$$

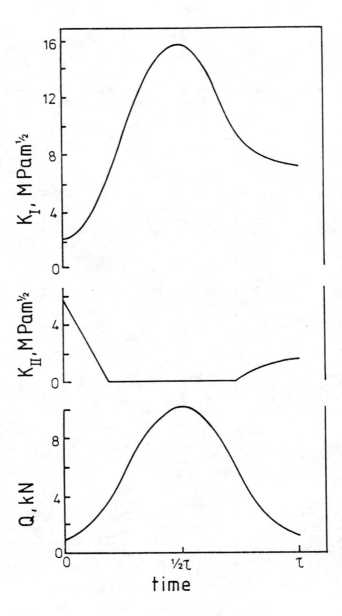

Fig. 6. Cyclic variation of stress intensity factors.

Equation 13 relates the dependence of $\Delta(CTOD)_c$ to the intrinsic tip stress intensity at threshold ΔK_{th}^i, σ_y' the cyclic yield stress and E, Young's Modulus. If the concept is correct, then ΔK_{th}^i should be proportional to $(E\sigma_y')^{\frac{1}{2}}$. The result in Figure 7 indeed shows such a relationship to hold for a range of metals and alloys which exhibit a dominantly transgranular mode of separation in the threshold region.

The above view is a global one indicating that the intrinsic contribution to fcg resistance improves with rigidity and yield strength of the matrix. Superimposed upon this is the detailed response of the material in the process zone ahead of the crack tip. The factors which can influence this behaviour include, slip character and crack branching, grain size, texture and environmental sensitivity.

The factors which influence the intrinsic contribution can in part also influence the extrinsic closure contribution. To illustrate this further, the influence of a number of microstructural variables on ΔK_{th} will be examined.

Microstructural Features Influencing ΔK_{th}

The influence of heat treatment and environment on the fcg response of an Al-Zn-Mg

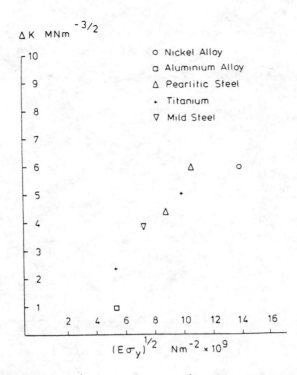

Fig. 7. A plot of ΔK_{th}^{i} versus $(E\sigma_y)^{\frac{1}{2}}$ for a range of metals and alloys

alloy is illustrated in Figure 8 (Heikkenen, 1982). The underaged material with a predominantly planar reversible slip character exhibited greater fcg resistance than the overaged alloy which exhibited multiple slip activity. The planar slip character- istics in the underaged alloy can lead to crack branching and mixed Mode I and II crack advance. The combination of slip reversibility, reduced effective stress

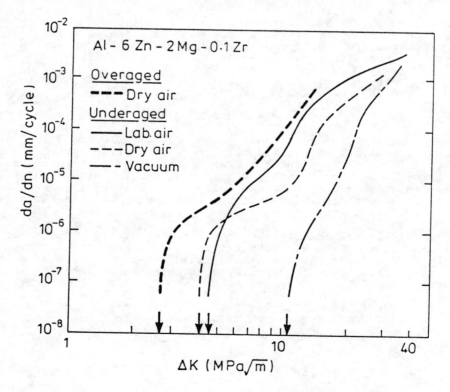

Fig. 8. The influence of ageing condition and environment on the fcg response of an AlZnMg alloy (Heikkenen et al, 1982).

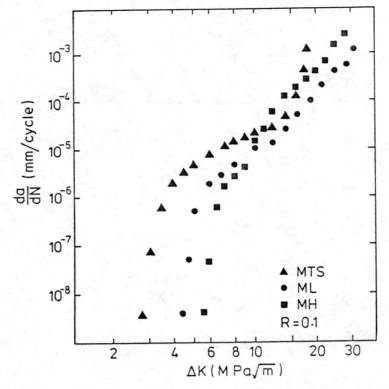

Fig. 9. The fcg response of an Al/Si/Mg alloy with Mn additions.
The alloy MH contains the highest volume fraction of
Al_2Mn_3Si dispersoids (Edwards and Martin, 1981).

intensity, the absence of environmental embrittlement in the process zones and a
significant closure contribution from the irregular fracture path lead to a substantially enhanced threshold value of $\sim$ 10 MPam$^{\frac{1}{2}}$.

The influence of dispersoids on the fcg response of Al/Mg/Si alloys with Mn additions
is illustrated in Figure 9 (Edwards and Martin, 1981). The alloy MH has the highest
volume fraction of Al_2Mn_3Si dispersoids. A combination of changes of slip character
and closure contribution can probably account for the increased fcg resistance with
increased volume fraction of dispersoids.

Grain size has been shown to have a marked influence on ΔK_{th} and its dependence has
been expressed in the form:

$$\Delta K_{th} = \Delta K_{th}(o) + kd^{\frac{1}{2}} \tag{14}$$

where $\Delta K_{th}(o)$ is ΔK_{th} for a notional grain size of zero and d is the grain size
(Carlson and Beevers, to be published). The results from tests on a stainless steel
are shown in Figure 10; ΔK_{th} increases with increasing grain size and decreasing R
ratio. This behaviour can probably be rationalized in terms of an increase in ΔK_{th}
with increasing grain size due to the formation of larger ligaments behind the
crack front and also to an increasing contribution to ΔK_{th}^C as the extent of irregular
crack growth increases with grain size.

A further example of the improvement in fcg resistance is illustrated in Figure 11
for a Ti-6Al-4V alloy, where a β heat treatment coarsened the α grain size from 12
to 40 μm (Halliday, 1981). The coarse grained material exhibited a greater resistance to fcg and this could be directly attributed to an increase in the closure
contribution to ΔK.

The information presented in Figures 7 to 11 shows that while a general trend in fcg
resistance can be established from a CTOD viewpoint the role of microstructural and
environmental features can dominate the overall material response.

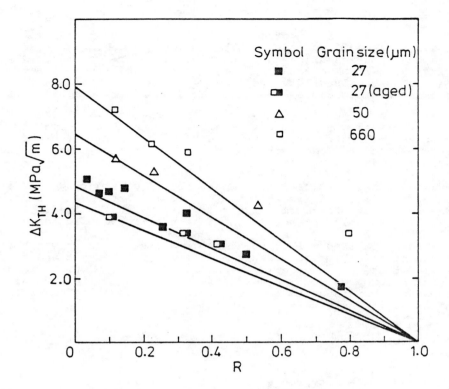

Fig. 10. The influence of R ratio on the fcg characteristics of a stainless steel for varying grain size (Priddle, 1978).

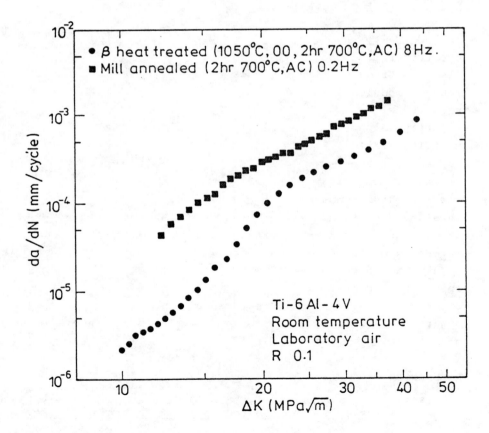

Fig. 11. The influence of β heat treatment on the fcg response of a Ti-6Al-4V alloy (Suresh, 1984).

CONCLUDING REMARKS

The resistance to fcg at low stress intensities is dominated by the magnitude of the fatigue threshold ΔK_{th} which can be thought of in terms of an intrinsic material resistance contribution ΔK_{th}^i and an extrinsic closure component ΔK_{th}^c. The factors influencing the two components are not mutually exclusive. For example environmental factors and slip character may reduce ΔK_{th}^i while the increased surface roughness leads to enchanced asperities and a larger ΔK_{th}^c contribution.

Recent work on the mode I and the mixed mode asperity model reveals that both are useful in elucidating non-closure behaviour. The mode I model provides a basis for focusing on the microstructural parameters which are related to closure interference. It can also be used to reveal the effects of asperity welding on the effective range of the stress intensity factor.

The mixed mode model discloses the importance of considering the local interference loading near the crack tip as well as the global, external loading. Both components of loading contribute to a complex stress state at the crack tip, and combine to produce conditions which promote branching on a microscopic scale.

In general terms it is possible to highlight those features which enhance ΔK_{th} and fcg resistance, namely, high elastic moduli for both matrix and fatigue fracture surface asperities, large asperities near the crack tip, high yield strength for the material in the process zone ahead of the crack tip. Slip reversibility, crack branching and a resistance to active environments can also be critical factors in determining the fcg resistance of metals and alloys.

SYMBOLS

D	Constant
ΔK	stress intensity range
ΔK_{th}	stress intensity range threshold value for fatigue crack growth
n	power exponent
ΔK_{th}^i	intrinsic threshold stress intensity
ΔK_{th}^c	crack closure contribution to ΔK_{th}
K_{cl}	the stress intensity when the global $K_I = 0$
K_{op}	the stress intensity to sustain an open crack with the faces touching but not transfering load
C	distance from asperity to crack tip
E	Young's modulus
b	asperity width
L	asperity height
G	rigidity modulus
ν	Poisson's ratio
R_p	plastic zone radius
$x_2 u_2$	vertical displacements
$x_1 u_1$	in plane displacements
$K_{(global)}$	stress intensity from external force
$K_{(local)}$	stress intensity from local force
P	local crack face force
B	specimen thickness
$\Delta CTOD$	crack tip opening displacement range
σ_y'	cyclic yield stress
d	grain size.

REFERENCES

Beevers, C J (1977) Some aspects of fatigue crack growth in metals and alloys. In D M R Taplin (Ed.), _Advances in Research on the Strength and Fracture of Materials_, Pergamon, Oxford, pp 239-260

Beevers, C J and co-workers (to be published) _Engineering Fracture Mechanics_

Carlson, R L and C J Beevers (to be published) _Engineering Fracture Mechanics_

Cooke, R J, P E Irving, G S Booth and C J Beevers, The slow fatigue crack growth and threshold behaviour of a medium carbon alloy steel in air and vacuum, Engineering Fracture Mechanics, 7, 69-77

Edwards, L and J W Martin (1981) The influence of dispersoids on fatigue crack propagation in Al-Mg-Si alloys, In D Francois (Ed.) Advances in Fracture Research, 1, 323-336

Halliday, M D and C J Beevers (1981), Some aspects of fatigue crack closure in two contrasting titanium alloys, J. Test Eval., 9, 195-201

Heikkenen, H G and co-workers (1982) The effect of environment on the low cycle fatigue and fatigue crack growth of a AlZnMg alloy, Georgie Institute of Technology Report

Kassir, M K and G C Sih (1973) Application of Papkovik-Neuber Potentials to a crack problem, Int. J. Solids Structures, 9, 643-654

Kendall, J M and J F Knott (1984) The influence of microstructure and temperature on near-threshold fatigue crack growth in air and vacuum, In C J Beevers (Ed.), Fatigue 84, EMAS, Warley, Birmingham, 1, 307-317

Meade, K P and L M Keer (1984) On the problem of a pair of point forces applied to the faces of a semi-infinite plane crack, J. Elast., 14, 3-14

Priddle, K E (1978) The influence of grain size on threshold stress intensity for fatigue cracks in AISI 316 stainless steel, Scripta Metallurgica, 12, 49-56

Suresh, S (1983) Crack deflection: Implications for the growth of long and short fatigue cracks, Metall. Trans. A, 14A, 2375-2385

Suresh, S (1984) Models for fatigue crack deflection, In C J Beevers (Ed.), Fatigue 84, EMAS, Warley, Birmingham, 1, 555-563

Suresh, S, G F Zamiski and R O Ritchie (1981), Oxide-induced crack closure: An explanation for near-threshold corrosion fatigue crack growth behaviour, Metall. Trans. A, 12A, 1435-1443

Walker, N J and C J Beevers (1980) A fatigue crack closure mechanism in titanium, Fatigue of Eng. Mats. and Struct., 1, 135-148

Ward-Close, C M and C J Beevers (1980) Influence of grain orientation on the mode and rate of fatigue crack growth in α-titanium, Metall. Trans. A, 11A, 1007-1017

Fatigue of
Non-Metallic Materials

R. J. YOUNG

Department of Materials, Queen Mary College, London, UK

ABSTRACT

The fatigue failure of a wide range of non-metallic materials has been reviewed. The
analysis of fatigue in such materials using S-N curves is described and the extent to
which fatigue crack propagation has been analysed using a fracture mechanics approach
is discussed.

KEYWORDS

Fatigue; fatigue crack propagation; polymers; bone; bone cements; wood; ceramics;
composites; ice.

INTRODUCTION

Non-metallic materials are being used increasingly in structural engineering applications although materials such as wood and bone have been used structurally for
millions of years. In discussion of their fatigue behaviour it has been necessary
to break the subject down into arbitrary classifications such as polymers, biomaterials, ceramics and composites, although there is bound to be overlap between the
different areas and features common to the different types of materials. Non-
metallics exhibit a wide range of fatigue behaviour. For example, polymers are
particularly prone to failure during cyclic loading and may also undergo 'thermal
fatigue' due to hysteretic heating. In contrast, ceramics are generally resistant
to fatigue failure. The different materials have also been the subject of widely
different levels of attention over the years. There is now a large and mature body
of literature concerning the fatigue of polymers. There is an increasing amount of
interest in fatigue in bone, but very little is published on the fatigue of such an
important structural material as wood.

The review is also concerned with the application of fracture mechanics to fatigue
in non-metallic materials. This has tended to lag behind similar studies on metals
but over the last 30 years considerable advances have been made. However, great care
must be exercised in this area as the deformation of many non-metallic materials is
not linear elastic and materials such as polymers exhibit viscoelasticity. Wood and
bone also have anisotropic mechanical properties. Both the limitations and successes
of the fracture mechanics approach are therefore detailed below.

POLYMERS

Polymers are a wide range of macromolecular materials which include plastics, rubbers
and adhesives and although the fatigue behaviour of the different types of polymers
varies in detail there are several features common to the fatigue of all polymeric
materials. There is now a wide body of literature concerning the fatigue behaviour

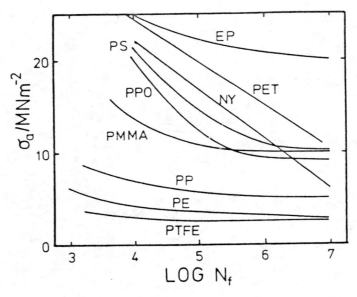

Fig. 1: S-N curves for several rigid polymers. After Kinloch
and Young (1983) using data of Riddell. See Table 1
for abbreviations

of polymers and several books and reviews have been published on the topic (Kinloch
and Young, 1983; Hertzberg and Manson, 1980; Radon, 1980; Williams, 1981; Sauer and
Richardson, 1980). Most of the original studies were concerned with the determination
of S-N curves. More recently, however, crack propagation has been followed using a
fracture mechanics approach.

S-N Curves

Much of the original work upon the fatigue of rigid polymers involved the generation
of S-N curves as can be seen in Fig. 1 where the logarithm of N_f, the number of
cycles to failure is plotted against the stress amplitude, σ_a. This approach has
given considerable useful information concerning the fatigue life of polymeric components but it does not generally provide much of an insight into the mechanisms of
the fatigue processes. However, one aspect of the behaviour which can be elucidated
from S-N curves is 'thermal' fatigue failure (Kinloch and Young, 1983). This arises
in polymers because of the viscoelastic nature of their deformation.

A perfectly elastic material will remain at the ambient temperature during a dynamic
fatigue test as no energy is dissipated in the specimen. However, polymers are
viscoelastic materials and the mechanical hysteresis they exhibit leads to energy
being dissipated as heat during each cycle. As they also have low thermal conducti-
vities their temperature can rise relatively quickly to high values above the melting
temperature or glass transition temperature of the polymer. The temperature rise and
rate of rise will depend upon a variety of factors which include testing variables
such as specimen geometry, test temperature, applied frequency and stress amplitude.
It also depends upon material properties such as the thermal conductivity and specific
heat capacity of the polymer and its viscoelastic properties. It is found therefore
that, unlike metals, thermal fatigue is a major failure mechanism in polymers.
Failure, however, does not necessarily take place through specimen fracture but it
can also occur by excessive plastic deformation taking place.

Examples of thermal fatigue in polymers can be quite spectacular. Sauer and Richardson
(1980) have pointed out that it is possible to run fatigue tests on polystyrene at
about 30 Hz and maximum stresses of about 15 MPa with a temperature rise of less than
2 K. In contrast, it is found that if poly (methyl methacrylate) is tested under
similar conditions it rapidly melts. This is because, even though the two polymers
have similar mechanical properties, such as Young's modulus and yield stress, poly
(methyl methacrylate) has a large viscoelastic damping maximum around room temperature
which is not present in polystyrene.

Crack Propagation

The kinetics and mechanisms of fatigue crack propagation in rigid polymers has
received considerable attention (Hertzberg and Manson, 1980). It is found that for
many such polymers log-log plots of crack growth rate da/dN versus the stress

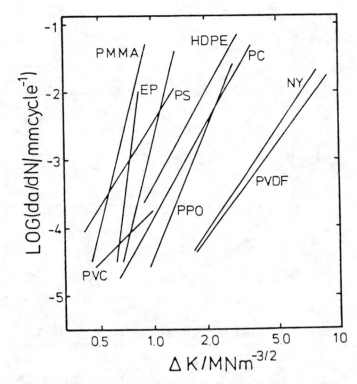

Fig. 2: Fatigue crack propagation data for several rigid polymers.
Data taken from Hertzberg and Manson (1980). See Table 1
for abbreviations.

intensity factor range ΔK generally give straight lines as can be seen in Fig. 2.
This shows that, as in the case of metals, these rigid polymers obey the Paris law
($da/dN \propto \Delta K^n$). The above approach requires the polymers to obey the assumptions of
linear elastic fracture mechanics (LEFM) (Kinloch and Young, 1983) and this is
generally found to be the case for many rigid amorphous and semicrystalline polymers.
For more flexible polymers such as rubbers or low density polyethylene which do not
obey the assumptions of LEFM plots are generally made of da/dN versus the fracture
energy range ΔG. Lake and Lindley (1966) showed that there was a complex relationship
between da/dN and ΔG for the dynamic fatigue of natural rubber. However, they went
on to show that the fatigue crack propagation data could be used to accurately predict
fatigue lifetimes of rubber components assuming that failure took place by the growth
of fatigue cracks from small intrinsic flaws. More recently, Teh, White and Andrews
(1979) have shown that a similar approach using ΔG can be used to follow fatigue crack
propagation in low density polyethylene. The approach has also found use in the
analysis of the fatigue of adhesive joints (Hertzberg and Manson, 1980). The mechanisms
of fatigue crack propagation in rigid polymers have received considerable attention
over recent years. It is thought to take place by the accumulation of damage in the
region ahead of the crack tip and this may be in the form of a localized shear
yielded zone, a single craze or a bunch of crazes, depending upon the type of polymer
and the experimental conditions (Kinloch and Young, 1983).

There has also been considerable interest in determining how changing experimental
variables such as test frequency, temperature and mean stress affect crack propagation.
Unfortunately no clear picture has emerged. For example, Hertzberg and Manson (1980)
have shown that increasing test frequency over the range 0.1 to 100 Hz decreases
fatigue crack growth rates for polycarbonate and nylon 66. Also increases in test
temperature may cause either increases or decreases in fatigue crack growth rates in
different polymers.

Attempts to find general correlations between fatigue crack growth in polymers and
fracture behaviour for monotonic loading have proved to be more successful as can be
seen from Fig. 3.

Hertzberg and Manson (1980) showed that there was a correlation between ΔK_r (the
value of ΔK required to drive a crack at a constant propagation rate) and the value
of K_{IC} for monotonic loading. It can be seen from Fig. 3 that the fatigue resistance
of the polymer increases as the value of K_{IC} rises. This behaviour is consistent
with the two-stage Dugdale model proposed by Williams (1977) to explain fatigue crack
propagation in rigid polymers.

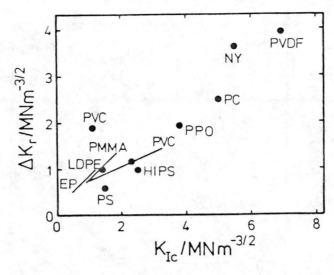

Fig. 3: Relationship between ΔK_r (at a constant value of da/dN of
7.5 x 10^{-4} mm cycle^{-1}) and K_{IC} for fracture under monotonic
loading. After Hertzberg and Manson (1980).

Kinloch and Young (1983) pointed out that care must be taken in using ΔK versus da/dN
data in the prediction of fatigue lifetimes of polymer components. They preclude the
important thermal failure mechanism common in polymers and also do not give any
information concerning the time required for crack initiation. This may account for
differences in ranking between Fig. 1 (unnotched specimens) and Figs. 2 and 3 (pre-
cracked specimens). Hence, although fatigue crack propagation takes place relatively
rapidly in epoxy resins their good performance in S-N experiments indicates their
resistance to crack initiation and thermal fatigue.

TABLE 1: Abbreviations used for Polymers

Abbreviation	Polymer Name
PMMA	Poly (methyl methacrylate)
PS	Polystyrene
PPO	Poly (phenylene oxide)
PE	Polyethylene
PTFE	Polytetrafluoroethylene
PP	Polypropylene
PE	Poly (ethylene terephthalate)
PVC	Poly (vinyl chlcride)
PVDF	Poly (vinylidene fluoride)
PC	Polycarbonate
EP	Epoxy resin
NY	Nylon 66
HIPS	High-impact polystyrene
LDPE	Low-density polyethylene
HDPF	High-density polyethylene

BIOMATERIALS

There is a rapidly increasing interest in the study of structure/property relationships
in biological materials which are used in structural applications such as bone and
wood. There are also important developments in the use of prosthetics for the replacement of natural materials in the body. This is all now described under the umbrella
title of 'Biomaterials'. As the subject has developed and matured within Materials
Science, the question of fatigue failure in these materials has naturally arisen.

Bone

The fatigue fracture of bone has been recognised clinically for more than a century
(Morris and Blickenstaff, 1967) as taking place in skeletal bone subjected to cyclic

loading. During a normal day the body is subjected to repeated or cyclic stressing but the magnitude of the stresses is normally less than that which will cause damage. Prolonged or intensive exercise such as marching or long-distance running has led to damage on a microscopic level but live bone has the remarkable ability to remodel itself and repair the damage. However, if the damage accumulates more rapidly than it can be repaired then fatigue failure may results (Carter and Hayes, 1976). This type of failure is sometimes called rather misleadingly "stress fracture" and has been documented as occuring in young military recruits ("march" fractures), athletes and ballet dancers (Wright and Hayes, 1976). It is reported that patients are often unaware of the developing fatigue crack until it is sufficiently large that it causes pain or even complete fracture of the bone occurs. It can take place in a wide variety of bones in the legs and feet (Morris and Blickenstaff, 1967).

The first study of fatigue in compact bone was reported by Evans and Below (1957) who measured fatigue lifetimes for rectangular cross-sectioned samples of human femur, tibia and fibula deformed in flexure. This was followed by further work by Evans and Riolo (1970) who correlated the variability of the fatigue lifetimes for a particular level of stress amplitude with the structure of the bone. As might be expected these measurements of fatigue lifetimes at a single stress amplitude have been followed by the generation of S-N curves. Carter and Hayes (1976) performed a detailed and careful study of the effect of specimen temperature and density upon S-N curves obtained from rotating cantilever specimens of compact adult bovine bone. Figure 4 shows S-N curves for samples of bovine bone of similar density deformed at two different temperatures. It can be seen that the fatigue lifetime of the bone material decreases as the temperature is increased and this has important practical implications. Although the normal body temperature of humans is 37°C significant variations can occur in limbs. For example, surface temperatures of feet of as low as 15°C have been measured in cold environments and of as high as 43°C for subjects walking in warm environments (Carter and Hayes, 1976). Since it is known that the temperature of deep tissue is only a couple of degrees different from that at the surface it is anticipated that the temperature of bones in feet could easily vary from 20°C to 40°C. The deterioration in fatigue performance with increasing temperature could well be a factor in causing "march" fractures and these observations could be significant for the design of military footware.

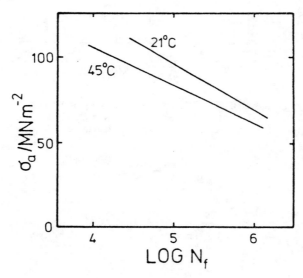

Fig. 4: S-N curves for samples of bovine femur tested at 21°C and 45°C. After Carter and Hayes (1976)

Carter and Hayes (1976) also showed that the fatigue lifetime of bovine bone was strongly dependent upon the density of the bone. For example, they reported a three-fold increase in fatigue life for just a 6% increase in density from 1860 to 1980 kg/m³. Again this observation has important practical implications especially for the elderly who sometimes suffer from fatigue fractures in the necks of their femurs. It has been found that there can be significant reductions (up to 37%) in the density of the material in the femoral necks of elderly patients and so it is anticipated that this would make this region prone to fatigue failure.

Carter and Hayes (1977) went on to investigate the mechanisms of fatigue failure in compact bone. They showed that there was a progressive loss of stiffness and strength during the lifetime of the specimens. Althought this is quite different from the behaviour of metals they noted that it is similar to that of fibre-reinforced composites (see later) which have structures similar to bone. They suggested that the behaviour was due to cumulative microcracking, debonding and fibre breakage in the bone.

It should be pointed out that the fatigue lifetimes determined from tests on dead bone will be conservative estimates when they are related to the performance of live bone. The natural healing response of bone in the body will tend to repair the damage and give longer lifetimes than for dead bone. Hence the S-N curves in Fig. 4 should be considered as lower bounds when predicting the behaviour of live bone.

It is only relatively recently that attempts have been made to apply fracture mechanics to the fatigue failure of compact bone. Wright and Hayes (1976) employed a double cantilever beam geometry to follow the growth of fatigue cracks parallel to the long axes of adult bovine femora and tibia. Typical results for da/dN versus ΔK are shown in Fig. 5 and it can be seen that the data obey the Paris law with exponents of the order of 3-5. They also showed that fracture took place by the propagation of cracks along the weak interfaces in the fibrous structure of the bone.

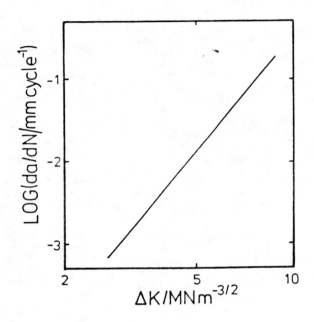

Fig. 5: Typical fatigue crack propagation data for bovine bone specimens tested with cracks parallel to the long axis of the bone. After Wright and Hayes (1976)

The extent to which the fracture mechanics approach outlined above is of use in the prediction of the behaviour of bone in service is questionable. For example, research has so far only been concerned with crack propagation parallel to the long axis but many fractures take place by cracks growing perpendicular to this axis. However, Bonfield and Behiri (private communication) have reported that it is difficult to propagate cracks using fracture mechanics specimens perpendicular to the long axis in bone. Nevertheless, considerable improvements in our understanding of this important material have been made and it is likely that because of the current levels of interest, we will see further advances in the near future.

Bone Cements

PMMA-based acrylic bone cements are now used widely in orthopaedic surgery as non-metallic implant materials. Since the implant is subject to mechanical loading when the patient recovers and becomes mobile there is consequently considerable interest in the mechanical properties of bone cements. The materials currently in use are developed from acrylic dental materials and the surgical grades of PMMA are self curing and supplied in the form of a powder and liquid monomer (Charnley, 1970). They are used widely for the fixation of prosthesis to bone in the reconstructive surgery of joints such as hips, knees and ankles. They have also found use in the fixation of fractures and the repair of bone defects. The use in such applications means that the cements are invariably subjected to considerable stresses.

Although such stresses are invariably cyclic in nature most of the interest in mechanical properties has been limited to monotonic loadings (Saha and Pal, 1984). It is generally found that bone cements are considerably weaker than both pure PMMA and compact bone and so considerable effort has been expended in the strengthening of bone cements. The main reason for the strengths being lower than normal grades of PMMA such as 'perspex' or 'plexiglass' is that the hand mixing means that bubbles and even body fluids become entrapped producing points of weakness in the structure. Also certain grades have additives such as $BaSO_4$ (to render them radio-opaque) and antibiotics both of which may affect mechanical properties.

There have been a few reports of studies of the fatigue behaviour of acrylic bone cements. Freitag and Cannon (1977) investigated the behaviour of two grades of bone cement using rotating bending fatigue apparatus. They performed measurements both in air at 22°C and in bovine serum held at 37°C (body temperature) and also investigated the effect of including different amounts of $BaSO_4$. Typical S-N curves for two different bone cements deformed in air are given in Fig. 6.

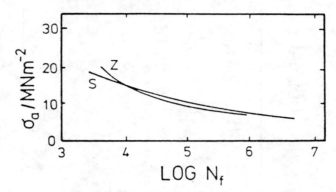

Fig. 6: S-N curves for Zimmer (Z) and Simplex-P (S) acrylic bone cements tested in air at 22°C. After Freitag and Cannon (1977)

Freitag and Cannon (1977) also found that in general the fatigue behaviour of the specimens tested in bovine serum at 37°C was superior to that of specimens tested in air at 20°C. In addition, they noted that the addition of up to 10% of $BaSO_4$ did not have a strong effect upon fatigue lifetimes.

More recently Wright and Robinson (1982) have applied fracture mechanics fatigue crack propagation in bone cements. They suggested that since there are numerous sources of stress concentration in bone cements in the body due to jagged-bone/bone cement interfaces and voids and seems within the cement caused by the mixing process it is likely that fatigue lifetimes are likely to be controlled by crack propagation rather than initiation. Hence they determined plots of da/dN versus ΔK for a series of different bone cements including one reinforced with short-strand carbon fibres, as can be seen in Fig. 7.

The behaviour of the normal bone cement is similar to pure PMMA but it can be seen that the addition of carbon fibres significantly improves the fatigue resistance of the cement. However, Wright and Robinson (1982) showed that when the data in Fig. 7 were normalised with respect to the elastic modulus of the material by plotting da/dN versus $\Delta K/E$ they all fell on the same line. They took this as an indication that fatigue crack propagation in these materials was strain controlled.

Further work in this area is needed to determine the fatigue behaviour of cement/bone or cement/metal interfaces. Hopefully this may allow the calculation of fatigue lifetimes for in-body applications. Obviously the failure of orthopaedic implants can lead to patients, who are often elderly, having to suffer the trauma of further major surgery. Any improvements in the materials and better information concerning their performance is of great interest to orthopaedic surgeons and of benefit to their patients.

Wood

Even though wood is on a volume basis one of the most widely-used structural materials, its fatigue behaviour has in the main been ignored. In fact, it was generally thought that wood did not suffer from fatigue failure and considerable experience with wooden aircraft structures during World War II gave rise to no major fatigue problems (Kyanka, 1980). However, in many applications of wood, conservative design criteria are used and so the material is not normally pushed to its limit. Wood is a biological material and subjected to continual cyclic deformation during its growth period from wind loading. This gave early workers confidence in the ability of the material to stand fatigue as it was thought to be 'fatigue conditioned' during growth (Kyanka, 1980).

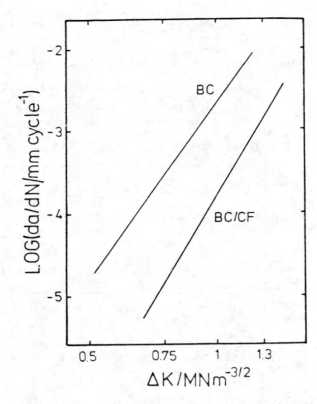

Fig. 7: Typical fatigue crack propagation data for an acrylic bone
cement (BC) and the same cement reinforced with short-
strand carbon fibres (BC/CF). After Wright and Robinson
(1982)

Wood is employed structurally in a wide variety of forms which include solid sections,
thin veneers in laminated structures and as chips or flakes in board products. It is
now recognised that fatigue can occur (Kyanka, 1980) and wood has definite endurance
limits giving S-N curves as can be seen in Fig. 8. It can be seen that the worst
behaviour is displayed by the particle board probably due to the presence of wood/
adhesive interfaces. It should be noted that it has been reported by Kyanka (1980)
that the fracture surfaces of wood broken in static tensile tests are indistinguishable
from those of wood which have failed by fatigue.

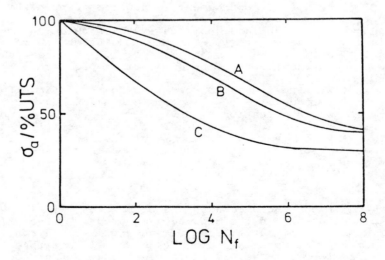

Fig. 8: Typical S-N curves for wood and wood composites,
A - Douglas fir (laminated) B - White oak (laminated),
C - Particle board. The stress amplitude is expressed
in terms of the ultimate tensile strength. After Kyanka
(1980)

Barrett (1981) has discussed the application of fracture mechanics to the propagation of cracks both parallel and perpendicular to the grain during the static loading of wood but as yet there does not appear to be any reports of da/dN versus ΔK curves being generated. Taking the analogy of the behaviour of bone and polymers it should be possible to propagate fatigue cracks at least parallel to the grain especially since wood is known to be a viscoelastic material and also Barrett (1981) has reported that slow crack growth and delayed fracture can occur in the material.

It would seem that there is considerable scope for the application of fracture mechanics to the fatigue of wood as many imponderables still exist. For example, different types of wood have different values of K_{IC} (Barrett, 1981). Also certain species of wood have been known for many years to be more durable during cyclic deformation than others (Kynaka, 1980). Oak and spruce were widely used in the structures of sailing ships and laburnum and yew were found to produce the most durable long bows. It is of considerable interest to determine what structural factors are responsible for the variation in fatigue performance of different species of wood.

COMPOSITES

The term 'composites' covers a wide range of multiphase engineering materials. It is normally used to describe systems which consist of a relatively weak matrix material reinforced with stiff, strong fibres and it is there materials that will be considered in this section. The application of fracture mechanics to fibre-reinforced composites subjected to monotonic loading is in its infancy and so there have been only a few reports of the analysis of fatigue in these materials using LEFM. Most of the work has involved the reporting of S-N data. Before the fatigue behaviour of composites is considered it is necessary to examine the fatigue of the fibres that are used as reinforcement.

Reinforcing Fibres

High performance reinforcing fibres such as glass, carbon and boron are generally relatively resistant to fatigue failure during cyclic deformation. In contrast polymeric fibres such as nylon, polyester and polyacrylonitrile are found to undergo fatigue failure when deformed to loads in excess of about 60% of their tensile strengths (Bunsell and Hearle, 1974). This is a particular problem when they are used as reinforcement in pneumatic tyres where fatigue of the cord fibres can lead to breakdown of the whole structure (Prevorsek and Lyons, 1971).

Very recently polymeric reinforcing fibres have been developed which are made from aromatic polyamides (aramids) and spun from liquid crystalline solutions of the polymer. The best-known example has the trade-name 'Kevlar' which has extremely high values of stiffness and strength (Magat, 1980) and is finding increasing use as a reinforcement for brittle polymeric matrices. Konopasek and Hearle (1977) have evaluated the fatigue behaviour of single aramid fibres and found that they were significantly more fatigue resistant than nylon or polyester fibres. Within the range of scatter within the data they found that cyclic deformation produced very little deterioration in strength compared with that determined by normal tensile tests.

Fibre-Reinforced Composites

The fatigue behaviour of fibre-reinforced plastics (Hertzberg and Manson, 1980; Hull, 1981; Owen, 1980) and metal matrix composites (Dvorak and Johnson, 1980) have been reviewed recently and so only the main points will be outlined here. Often, composite materials can appear to be very fatigue resistant since even in the notched condition their fatigue strengths can be 60-70% of their static strength (Reifsnider, 1980). However, they can undergo damage during cyclic loading at high stresses. This can lead to a significant reduction in stiffness which may lead to 'failure' in components where the stiffness is critical.

There have been a number of reports (Hull, 1981; Owen, 1980; Furue and Shimamura, 1980) of the generation of S-N curves for discontinuous-fibre composites consisting of short fibres in both thermoplastic and thermosetting matrices. Such materials are particularly useful as they can be fabricated relatively easily using conventional methods used with single-phase polymers such as injection or compression moulding. It is often difficult to compare data from different workers since an enormous variety of fibres (of different length), matrices, surface treatments and fibre volume fractions have been employed. However, Fig. 9 shows an example of typical behaviour and gives an indication of the mechanisms of fatigue damage in short-fibre composites.

It can be seen that significant levels of damage can occur at stresses well below those needed to cause complete separation of the two halves of the specimens. For

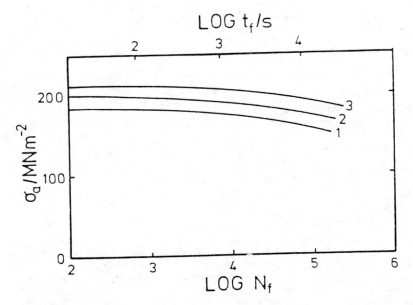

Fig. 12: S-N curves and static fatigue data for polycrystalline
alumina. Curve 1 is for cyclic fatigue and curves 2
and 3 are for static fatigue. After Guiu (1978).

ICE

The increase in the exploration of mineral resources in cold regions has led to an
awareness of the importance of gaining a better understanding of the mechanical
properties of ice. However, it is only recently that fatigue crack propagation has
been unambiguously demonstrated in this material. Nixon and Smith (1984) have
produced fatigue life data for cylindrical polycrystalline ice specimens made from
distilled water and subjected to reversed cyclic loading at -13°C. They looked at
two types of ice with different grain sizes and degrees of porosity and the data are
shown in Fig. 13.

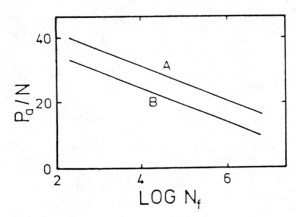

Fig. 13: Fatigue data for two types of ice deformed at -13°C at
2.8 Hz, A - porous, B - very porous. P_a = maximum load
amplitude. After Nixon and Smith (1984).

It can be seen that the samples with a high degree of porosity have fatigue lifetimes
about one order of magnitude less than the samples with less porosity.

Nixon and Smith also took replicas of the fracture surfaces and found striations
perpendicular to the crack growth direction at a spacing consistent with the amount
of known crack growth per cycle. They suggested that their results may be of significance to the understanding of the break up of ice in situations such as under the
action of cyclic wave loadings.

CONCLUSIONS

It is clear from the work described above that considerable fatigue data have been generated for non-metallic materials over the past 30 years. The bulk of this data is in the form of S-N curves. Some data for cyclic crack growth (da/dN v. ΔK) has been obtained, notably for polymers. This approach has not however reached the same stage of maturity as the similar approach for metals and its use in the design of non-metallics is limited. With their increasing use in critical structural applications it remains a considerable challenge for engineers and designers to avoid over-conservative design criteria and to use non-metallic materials to their full potential.

REFERENCES

Barrett, J D (1981). Fracture mechanics and the design of wood structures. Phil. Trans. Roy. Soc. Lond., A299, 217-226

Beaumont, P W R and B Harris (1971). The effect of environment on fatigue and crack propagation in carbon-fibre reinforced epoxy resin. Proceedings of the International Conference on Carbon Fibres, their Composites and Applications, paper 49, Plastics Institute, London

Bursell, A R and J W S Hearle (1974). The fatigue of synthetic polymeric fibres. J. Appl. Polym. Sci., 18, 267-291

Carter, D R and W C Hayes (1976). Fatigue life of compact bone - I. Effects of stress amplitude, temperature and density. J. Biomech., 9, 27-34

Carter, D R and W C Hayes (1977). Compact bone fatigue damage - I. Residual strength and stiffness. J. Biomech., 10, 325-337

Charnley, J (1980). Acrylic Bone Cement in Orthopaedic Surgery. Williams and Wilkins, Baltimore

Dvorak, G J and W S Johnson (1980). Fatigue of metal matrix composites. Int. J. Fract., 16, 585-607

Evans, A G (1980). Fatigue in ceramics. Int. J. Fract., 16, 485-498

Evans, F G and M Lebow (1957). Strength of human compact bone under repetitive loading. J. Appl. Physiol., 10, 127-130

Evans, F G and M L Riolo (1970). Relations between the fatigue life and histology of adult human cortical bone. J. Bone Joint Surgery, 52-A, 1579-1586

Freitag, T A and S L Cannon (1977). Fracture chracteristics of acrylic bone cements II. Fatigue. J. Biomed. Mater. Res., 11, 602-624

Furue, H and S Shimamura (1980). Fatigue behaviour of rigid polymeric materials and composites at room and elevated temperature. Int. J. Fract., 16, 553-561

Garrett, G C, H M Jennings and R B Tait (1979). The fatigue hardening behaviour of cement-based materials. J. Mater. Sci., 14, 296-306

Guiu, F (1978). Cyclic fatigue of polycrystalline alumina in direct push-pull. J. Mater. Sci., 13, 1357-1361

Hertzberg, R W and J A Manson (1980). Fatigue of Engineering Plastics. Academic Press, New York

Hull, D (1981). An Introduction to Composite Materials. Cambridge University Press, Cambridge

Kinloch, A J and R J Young (1983). Fracture Behaviour of Polymers. Applied Science, London

Konopasek, L and J W S Hearle (1977). The tensile fatigue behaviour of para-orientated aramid fibres and the fracture morphology. J. Appl. Polym. Sci., 21, 2791-2815

Kyanla, G H (1980). Fatigue properties of wood and wood composites. Int. J. Fract., 16, 609-616

Lake, G J and P B Lindley (1965). The mechanical fatigue limit for rubber. J. Appl. Polym. Sci., 9, 1233-1251

Magat, E E (1980). Fibres from extended chain aromatic polyamides. Phil. Trans. R. Soc. Lond., A294, 463-472

Morris, J M and L D Blickenstaff (1967). Fatigue Fractures. Thomas, Springfield, IL.

Nixon, W A and R A SMith (1984). Preliminary results on the fatigue behaviour of polycrystalline freshwater ice. Cold Reg. Sci. Tech., 9, 267-269

Owen, M J (1980). Fatigue processes in fibre reinforced plastics. Phil. Trans. Roy. Soc., A294, 535-543

Prevorsek, D C and W J Lyons (1971). Endurance of polymeric fibres in cyclic tension. Rubb. Chem. Tech., 44, 271-293

Radon, J C (1980). Fatigue crack growth in polymers. Int. J. Fract., 16, 533-552

Reifsnider, K (1980). Fatigue behaviour of composite materials. Int. J. Fract., 16, 563-583

Saha, S and S Pal (1984). Mechanical properties of bone cements: A review. J. Biomed. Mater. Res., 18, 435-462

Sauer, J A and G C Richardson (1980). Fatigue of polymers. Int. J. Fract., 6, 499-532

Smith, T R and M J Owen (1968). The progressive nature of fatigue damage in glass reinforced plastics. Proceedings of the 6th International Resins and Plastics Conference of the British Plastics Federation, Paper 27, BPF London

Teh, J W, J R White and E H Andrews (1979). Fatigue of viscoelastic polymers: I Crack-growth characteristics. Polymer, 20, 755-763

116

Williams, J G (1977). A Model for fatigue crack growth in polymers. J. Mater. Sci., 12, 2525-2533

Williams, J G (1981). Fracture mechanics of non-metallic materials. Phil. Trans. Roy. Soc. Lond., A299, 59-72

Wright, T M and W C Hayes (1976). The fracture mechanics of fatigue crack propagation in compact bone. J. Biomed. Mater. Res. Sym., 7, 637-648

Wright, T M and R P Robinson (1982). Fatigue crack propagation in poly (methyl methacrylate) bone cements. J. Mater. Sci., 17, 2463-2468

Damage Tolerant Design

J. D. HARRISON

The Welding Institute, Abington, UK

ABSTRACT

Most engineering structures contain crack or crack-like imperfections ab initio. Damage tolerant fatigue design ensures that these inherent cracks will not propagate to failure either within the design life or between inspection periods. This concept depends on the application of fracture mechanics to fatigue crack growth. There are a number of components required for a satisfactory damage tolerant design and great strides have been made in all these areas in research over the last 20 years.

Fatigue crack growth law

Crack growth data now exist for a variety of materials. Recent work has concentrated on the effects on this relationship of such variables as environment, cyclic frequency and R-ratio and also on the threshold for crack propagation, on which much pioneering work was carried out by Frost and his co-workers at NEL.

Cracked body stress analysis

Advances in numerical methods have improved both accuracy and computation speed. Numerical methods can be used to compare alternative designs, to study the effect of weld profile and thickness, etc.

Inspection

It is clear from fatigue crack propagation analysis that the period between in-service inspection should not be constant but should decrease with life.

The paper will discuss these developments in the context of the author's own experience concerned with welded stuctures, wherein an important development has been the publication of BS PD 6493, giving methods of assessing the significance of weld defects. Some recent case histories will be reviewed.

KEYWORDS

Fatigue, crack propagation, fracture mechanics, welding, fatigue threshold, weld imperfections, non-destructive testing.

INTRODUCTION

It is the author's understanding that "damage tolerance" was first introduced as a formal concept for the design of military aircraft. For such aircraft, it was impractical to adopt the fail-safe procedures used for civil aircraft. As an alternative it was required that the designer demonstrate that some assumed pre-existing defect would not propagate to failure between two inspections, the first being assumed to be made at a time when the defect was at the threshold of detectability.

However, the concept of damage tolerant design is much broader and in this paper, the author will concentrate on its application to welded structures.

It has for long been recognised that weldments (in common indeed with all metallic materials) are inherently flawed from completion. A feature, which distinguishes weldments (and castings) from wrought products, is that the initial imperfections are more randomly oriented and that they are often perpendicular to the direction of stress. (In wrought products the imperfections are usually parallel to the surface and therefore also usually parallel to the direction of stress). Furthermore, the weld profile introduces a concentration in stress, and finally, welds are often sited at regions of gross structural discontinuity, changes of thickness, penetrations, etc., which bring about additional stress concentration. For these reasons, welded structures are particularly prone to fatigue failure, and it has been estimated that over 90% of failures in welded components subjected to cyclic loading are attributable to this cause.

The fatigue process is normally looked on as comprising the two major phases, initiation and propagation. In the 1960s research was carried out into the reasons for the very disappointing fatigue behaviour of welded high strength steels. Whilst the fatigue strength of unnotched steel or of steel with a mild stress concentration, such as a hole, rises with increasing tensile strength, the fatigue strength of a fillet welded joint is completely unaffected by the tensile strength of the parent steel. This was found to be true for steels ranging in strength from mild steel, $\sigma_Y \approx 250 \text{N/mm}^2$, to maraging steel, $\sigma_Y \approx 1,750 \text{N/mm}^2$. This behaviour is shown schematically in Fig.1. These observations led Signes and co-workers (1967) to investigate the toe region of welded joints where fatigue cracks initiated. It was found by these workers and by Watkinson, Bodger and Harrison (1970) that, at the weld toe, there were invariably small sharp intrusions of welding slag, ranging in depth from ≈0.01 to 0.2mm. These were often located at the bottom of undercut having a typical depth of 0.2mm, so that the total depth to the bottom of the intrusion would approach 0.5mm. A typical section showing such an intrusion acting as a fatigue starter is shown in Fig.2. From these studies it became apparent that virtually the entire fatigue life of welded joints was taken up with crack propagation. A significant initiation phase would only be introduced if these inherent discontinuities were removed by full machining or by local grinding of the weld toe. In more recent work, Smith and Smith (1982) have confirmed that the initiation phase in the fatigue of fillet welded joints is of negligible duration. Using sophisticated potential drop measurements, they were able to detect crack growth after only 5% of the total life. However, possibly because of improved welding technique or quality control, the initiating weld toe discontinuities were considerably smaller than those observed by Watkinson and co-workers, with an average depth of 0.045mm and a 95% upper bound of 0.1mm. Smith and Smith also studied the root radii of the weld toe discontinuities and found them to average 2.6μm, with some root radii of less than 1μm. With discontinuities as sharp as this, it is little wonder that fatigue crack propagation started almost immediately.

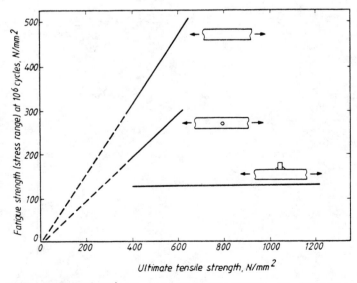

Fig.1. Affect of tensile strength on fatigue strength.

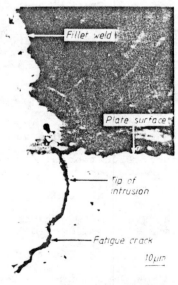

Fig.2. Section at fillet weld toe showing slag intrusion initiating a fatigue crack.

Signes and co-workers (1967) and Watkinson and co-workers (1970) concluded that the disappointing fatigue performance of welded high strength steels stemmed from the fact that fatigue cracks propagated at similar rates in these steels as in mild steel. This has since been confirmed in numerous fatigue crack propagation studies (e.g. Richards and Lindley, (1972)), which have showed that the rate of crack propagation is little affected by tensile strength. The high fatigue strength of unnotched high strength steels stems from their superior resistance to crack initiation.

The realisation of the overwhelming importance of crack propagation in determining the fatigue behaviour of welds fortunately coincided in time with the development of the application of fracture mechanics to fatigue crack propagation based on the early work of Paris and Erdogan (1963).

Paris' law is obviously very well known but will be repeated here for the sake of completeness.

$$\frac{da}{dN} = A(\Delta K)^n \tag{1}$$

ΔK can be expressed in terms of geometry, stress and crack size as

$$\Delta K = Y \, \Delta \sigma \, \sqrt{\pi a} \tag{2}$$

Where Y is a function of the geometry and of the instantaneous crack size, a.

The great power of fracture mechanics, as applied to fatigue, is that Eq.[2] can be substituted into Eq.[1], terms in "a" can be separated out and the equation can be integrated:-

$$\int_{a_1}^{a_2} \frac{da}{Y^n(\pi a)^n} = A(\Delta \sigma)^n \, N \tag{3}$$

If the integration is carried out over the entire range of crack size from the initial imperfection, a_i, to that at final failure, a_f, Eq.[3] represents the S-N curve for the particular geometry.

In order to apply this concept in practice, knowledge is needed about the three corners of the famous triangle of fracture mechanics; the crack growth law (values of A and n); the flaw sizes (a_i and a_f) and the stress range ($\Delta \sigma$). In addition, it is necessary to have available the cracked body stress analysis to give the parameter "Y" as a function of "a". These aspects are reviewed briefly below, followed by mention of three important codes which apply these concepts and finishing with some case histories, together with the implications with regard to inspection.

FATIGUE CRACK GROWTH

Other papers in the present volume (Knott (1985), Lindley and Nix (1985), and Beevers and Carlson (1985)) deal with a variety of aspects of fatigue crack growth. Here, therefore, we will concentrate on a few salient points.

da/dN versus ΔK relationship

Crack propagation data have been obtained for a great variety of materials under various environmental conditions.

An early collection of data was published by Frost, Pook and Denton (1971). This was a reanalysis in fracture mechanics terms of crack growth data obtained at the National Engineering Laboratory (NEL). Materials covered were a range of steels with tensile strengths from 430 to 2000N/mm^2, aluminium alloys, titanium and alloys, copper, nickel, Inconel, Monel, Phosphor bronze and brass; an impressive list. Maddox (1974a) concentrated on the generation of data relevant to the fatigue behaviour of welds in structural steels and obtained data for the parent steels, simulated heat affected zones and weld metals.

Richards and Lindley (1972) studied crack growth in a variety of ferrous alloys with a wide range of microstructures. Their work included investigations of the effects of R-ratio, thickness and test temperature. They noted that cracks could grow faster when there were local areas of cleavage or dimpled rupture interspersed with genuine fatigue cracking.

A general observation from all these data was that, for true fatigue crack growth, there is little effect of the strength of a given type of material (steel, aluminium alloy, etc.) on growth rates. Furthermore, it was also observed that there was little difference between materials if crack growth rate was expressed in terms of $\Delta K/E$ (Pearson, (1966), Pook and Frost (1973), Speidel (1982)). Speidel's data were obtained for materials ranging from lead to tungsten, representing a factor of 20 on Young's modulus.

These observations were of great importance with regard to the fatigue behaviour of welded structures. Firstly, as noted in the Introduction, they explained the poor fatigue performance of welded high strength steels. Secondly, they make it possible to derive design rules for welded joints in other alloy systems, based on the very large volume of data already established for steels, by simply factoring the stresses in the S-N curves in proportion to the Young's moduli.

Equation [1] implies that the relationship between da/dN and ΔK would be linear in a log/log plot. In fact, of course, the relationship is usually sigmoidal of the type shown in Fig.3. There is a threshold, ΔK_{th}, below which no crack growth occurs, whilst at the high ΔK end, the growth rate accelerates rapidly as the maximum value of K, K_{max}, approaches the stress intensity for fracture, K_c.

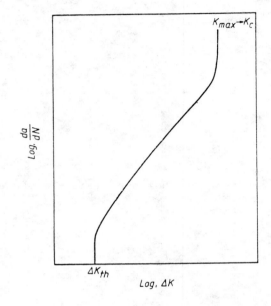

Fig.3. Idealised log da/dN v log ΔK relationship.

Much pioneering work on thresholds was done at the NEL by Frost and co-workers in their studies of non-propagating cracks. See, for example Frost (1959). In particular they studied the threshold in terms of the critical stress required to drive a crack. They proposed that, whether or not a crack would grow in a simple shallow notched bar loaded in cyclic tension, depended on whether or not $(\Delta \sigma)^3 a$ exceeded a critical value, C. Subsequently, both the present author (1970) and Frost, Pook and Denton (1971) reanalysed the NEL data using fracture mechanics methods and showed that they could be expressed in terms of a stress intensity factor threshold, ΔK_{th}. This differs only slightly from the original $(\Delta \sigma)^3 a$ relationship.

For many practical purposes, it is sufficient to express the total law in terms of Eq.[1] for $\Delta K > \Delta K_{th}$ and $da/dN = 0$ for $\Delta K < \Delta K_{th}$. The acceleration which occurs as $K_{max} \rightarrow K_c$, seldom has any significant bearing on the outcome of the analysis, because the majority of the life is spent when the crack is short and K_{max} is therefore relatively low. (As a very approximate rule of thumb, it can be said that about half of the remaining life will be spent in doubling the current crack size).

The above ignores the curvature of the log da/dN versus log ΔK plot between the threshold and the linear Paris relationship. For analytical purposes, greater accuracy can be achieved by representing the entire law by a series of, say, five straight lines.

Alternatively, empirical relationships have been fitted to the data. These have been reviewed by Austen (1979), who suggested a law of the form:-

$$\frac{da}{dN} = A(\Delta K)^n \left[\frac{(\Delta K - \Delta K_{th}) \cdot \Delta K \cdot K_c}{K_c - \Delta K/(1 - R)} \right]^p \qquad [4]$$

Many other similar laws have been proposed. A difficulty which arises with Eq.[4] is that there are five materials parameters, A, n, ΔK_{th}, K_c and p. In order to evaluate these with reasonable certainty, a large volume of data is required. However, values of A, m and p can be derived theoretically as:

$$A = \frac{1}{4\pi \ \sigma_{YS} \ E} \ ; \ n = 2 \text{ and } p = 0.25$$

Factors affecting the Crack Growth Relationship

For a given material, the crack growth relationship is affected in a synergistic way by a variety of factors; in particular, the R-ratio, the environment, and the cyclic frequency.

The R-ratio. Frost, Pook and Denton's work (1971) showed clearly that the R-ratio has a marked effect on the threshold, ΔK_{th}, with ΔK_{th} falling with increasing R. More recently, reviews of the effect of R on ΔK_{th} for steels have been carried out by Garwood (1979) and by Davenport (1982).

In general the empirical results have been expressed in the form:

$$\Delta K_{th} = \alpha + \beta(1-R)^{\gamma} \qquad [5]$$

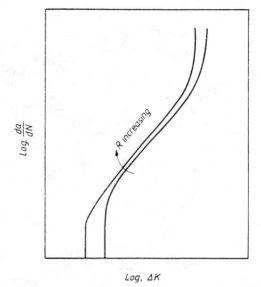

Garwood (1979) suggested that a reasonable lower bound for carbon and carbon manganese steels in air was given by

$$\Delta K_{th} = 46 + 144(1-R) \, Nmm^{-3/2} \qquad [6]$$

As well as affecting ΔK_{th}, increasing positive R-ratios also lead to higher growth rates in the finite growth rate régime, as implied by Eq.[4], although this effect is small for structural steels. The overall effect of R is shown schematically in Fig.4.

Fig.4. Effect of R-ratio.

Fatigue cracks are driven by tensile stresses. Thus, for negative values of R, all or part of the compressive portion of the cycle can be ignored.

This variation in behaviour with R-ratio has often been studied in terms of crack closure and its counterpart, the effective range in ΔK, ΔK_{eff}, during which the crack is open. An example where such an approach was used to study crack propagation with R-ratios of -2 to +0.5 and where these results were used to predict the behaviour of fillet welds in an aluminium-zinc-magnesium alloy, is reported by Maddox and Webber (1978).

In this context it should be noted that, for structural assessments, the R-ratio of importance is that actually occurring in the cracked region. Due to local residual stress, this may bear little relationship to the R-ratio for the external load. High tensile residual stresses (of yield magnitude) exist in welded joints in the as-welded condition. For these, the stress locally pulsates downwards from yield whatever the external R-ratio. Thus, appropriate crack growth rates and thresholds are those obtained at high R. Even after postweld heat treatment, tensile residual stresses of about $100 N/mm^2$ remain. Whilst this is low compared to the as-welded situation, it is still quite high compared to a typical fatigue stress range. Thus, even after stress relief, a negative R-ratio for the external load may still imply a positive R-ratio locally. On the other hand, some methods for improving fatigue strength, such as hammer or shot peening rely on the introduction of compressive residual stress, so that the local R-ratio is negative, even if externally it is positive.

Environment and frequency. Much research has been carried out into the effect of various environments on fatigue crack growth. The largest volumes of information exist for structural and ship hull steels in sea water, for nuclear pressure vessel steels in PWR and BWR high temperature water environments and for austenitic stainless steels in liquid sodium.

For conventional alloys at ambient temperature, frequency has little effect on fatigue behaviour. However, for corrosive environments, this is not the case. In general, the lower the frequency, the higher the crack growth rate. There is also a combined effect of the R-ratio, in that, in a corrosive environment, an increase in R-ratios leads to a larger increase in growth rate than in air.

Cathodic polarisation for steels in sea water can reduce the crack growth rate compared to that under free corrosion and may even restore it to that measured in air.

The threshold for cathodically polarised specimens in sea water may even be higher than that in air. This is attributed to the crack wedging action of calcareous deposits.

For further information on the effects of specific environments the reader is referred to the appropriate literature. Some environments are of particular interest. The effects of sea water in general, are discussed in the book by Jaske, Payer and Balint (1981) and, the effects of sea water on crack growth in offshore structural steels, by Walker (1981). The effects of PWR environments on pressure vessels and piping steels are discussed by Cullen and Torronen (1980) and those of liquid sodium on crack growth in austenitic stainless steels by James and Knecht (1975).

Summary. In conclusion, the practising engineer should usually be able to find crack growth data which are directly relevant to his particular situation. One note of warning should be sounded concerning the statistical treatment of crack growth data. This has not been well investigated. The user can derive an unwarranted confidence in the data because many da/dN v. ΔK data points can be established from a single test specimen.

THE DISCONTINUITY

Type and Size

The size of the initial discontinuity can obviously vary widely according to the material, product type, etc.

For welds, there are a variety of discontinuity types:

Planar	Cracks
	Lack of root side wall fusion, etc
Non planar	Inclusions
	Porosity

The planar discontinuities can be divided into those inherent to the welding process, as mentioned above, which occur in the toe region of all welds and are of the order 0.05 to 0.5mm deep, and discontinuities arising as a result of metallurgical or operator problems which can, of course, be much deeper. The fatigue behaviour of components containing planar discontinuities can be analysed on a crack propagation basis. For non-planar discontinuities there may be a significant initiation period, so that analysis in terms of crack propagation may be unduly conservative.

Except in exceptional circumstances, where the product is machined or ground, one simply has to live with the small inherent discontinuities present at all weld toes, and damage tolerant design must allow for them. It should also allow for larger discontinuities, since these may occur and escape detection by non-destructive testing (NDT). The certainty of finding a discontinuity of any given kind or size clearly varies both with the type of structure under consideration and with the NDT methods adopted. It is very difficult to answer the question "What size of defect might be missed?" The methods of NDT can be divided into those applied to the surface only (magnetic particle and dye penetrant) and the volumetric methods (radiography and ultrasonics). The surface methods are reliable for detection and surface length measurement but are incapable of measuring crack depth which is vital for a fracture mechanics assessment. Radiography is good at finding the relatively innocuous non-planar discontinuities, but is unreliable for detecting cracks except in thin materials and when these are favourably oriented. Ultrasonics is potentially the most satisfactory technique; but, when used conventionally with a single probe acting as both transmitter and receiver, it may be impossible to find large planar flaws, if these are perpendicular to the surface. This was highlighted in the first Plate Inspection Steering Committee Programme (PISC), where, in a round-robin study, a large defect was missed by a number of laboratories (Anon, 1979). Such problems are much less likely when more sophisticated tandem or multi-probe methods are used.

Assuming that a discontinuity has been found, it then has to be characterised and sized. Perhaps the main advantage of radiography is the relative ease with which weld discontinuities, once found, can be characterised. However, even if a crack is found by radiography, its through-thickness dimension cannot be measured. For this purpose ultrasonics is the only viable technique. However, the accuracy of defect sizing by ultrasonics is subject to wide variation. This was studied in a round robin programme involving the CEGB, the UKAEA Harwell and The Welding Institute (Jessop and co-workers, (1982)). It was found that, using conventional techniques, planar discontinuities were, on average, undersized by 3 to 4mm with 95% confidence limits of about +6 to -12mm. The most accurate method was that using the time-of-flight (Silk, (1979)) for which the 95% confidence limits were about +4 to -5mm. Clearly such limited sizing accuracy should be taken into account in any important assessment. It may swamp many other aspects of a defect tolerance analysis. Mudge and Williams (1981) discuss the implications of sizing inaccuracy with regard to the statistical aspects of the fracture mechanics analysis of the conditions for final instability.

Shape

The flaw shape has a bearing on the "Y" factor (see below). Flaw shapes vary both initially and during propagation. Maddox (1975) studied the variations in flaw shape at fillet weld toes for specimens loaded in tension. He found a rather widely scattered linear relationship between crack depth and crack length. Equations for the mean and upper and lower 95% confidence limits were as follows:-

$$2c = 13.6 + 2.58a \qquad \text{mm (upper 95\% confidence limit)} \qquad [7a]$$
$$2c = 6.71 + 2.58a \qquad \text{mm (mean)} \qquad [7b]$$
$$2c = 2.58a \qquad \text{mm (lower 95\% confidence limit)} \qquad [7c]$$

More recently other workers have studied this question. Smith and Smith (1982) carried out experimental measurements of the evolution of crack shapes at fillet weld toes for tension loading. They found a mean relationship of remarkably similar slope to that established by Maddox (1975)

$$2c = 2.65a \qquad [8]$$

Sommer, Hodulak and Kordisch (1977) argued theoretically that the stable condition for a semi-elliptical surface crack would be represented by

$$c/a = \left(\frac{M_{sc}}{M_{sa}}\right)^{\left(\frac{2n}{n+2}\right)} \qquad [9]$$

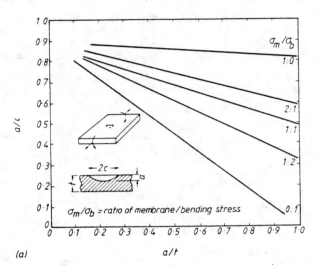

(a)

Where a is the depth and c the half-length of the crack, n is the exponent in the crack growth law and M_{sc} and M_{sa} are the stress intensity magnification factors for a surface crack, at the ends and bottom of the crack, respectively.

Iida (1980, 1983) and Clayton and Morgan (1984) have investigated the evolution of crack shape under tension, bending and combined bending and tension. Under bending loading short cracks tend to grow along the surface before developing through the thickness, whereas under tensile loading, relatively long shallow cracks will grow in the through-thickness direction initially. For any given loading condition there is a stable crack shape relationship towards which any given initial crack will tend. These findings are summarised in Fig.5.

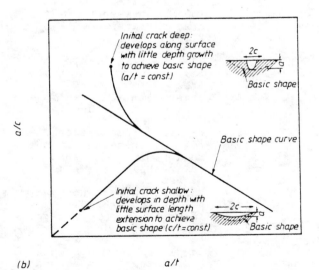

(b)

Fig.5. Basic crack shape curves for a semi-elliptic defect in a flat plate. (After Clayton and Morgan (1984)).

In-service inspection

In view of the uncertainty of defect detection already referred to, it is the author's view that in-service inspection should be relied on as a primary line of defence only in circumstances where sophisticated non-destructive testing techniques are employed. It is preferable to rely on redundancy/fail-safe design. In structures, redundancy is achieved by the provision of a number of parallel load paths. In pressure vessels etc., a form of redundancy is provided, as long as they have

124

adequate toughness to ensure leak-before-break. One alternative or supplement to redundancy is the use of some form of whole body monitoring which will detect the growth of cracks before they become critical. This can be provided by such techniques as flooded member detection for offshore structures (Bennett and Brown, (1983)) or, a relatively new technique which shows promise - acoustic pulsing (Bartle and Mudge, (1983)).

Where in-service inspection using sophisticated methods is considered as one of the primary lines of defence, fracture mechanics can be used to plan the frequency of inspections. This aspect is discussed by Williams and Mudge (1982), who also consider the accuracy required of the NDT technique for such assessments. The method proposed allows the assessment to be updated at each inspection.

STRESS

Residual Stress

The importance of residual stress has already been mentioned, affecting, as it does, the local R-ratio.

Stress Analysis of Cracks in Welded Components

Great strides have been made over the last 15 years in the development of methods for the stress analysis of cracked bodies, in particular those employing finite element methods.

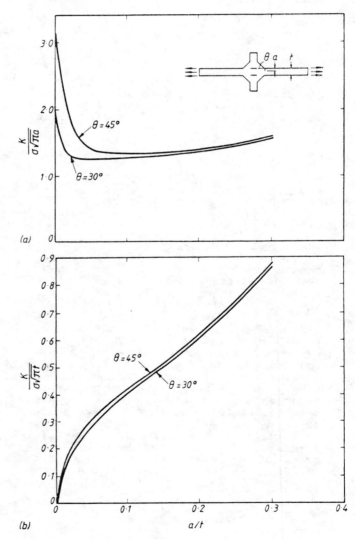

Maddox (1975) originally suggested that the stress intensity magnification factor, Y, for a crack growing at a fillet weld toe could be broken down into a number of components:

$$Y = \frac{M_s \, M_t \, M_k}{\phi_o} \qquad [10]$$

where M_s = Magnification factor due to the presence of the free surface
≈ 1.12.

M_t = Magnification factor due to the proximity of the free surface on the other side of the ligament ahead of the crack.

M_k = Magnification factor due to the presence of the weld. For very shallow cracks $M_k \rightarrow K_t$ the elastic stress concentration factor (SCF). However, M_k decays rapidly towards 1.0 for deeper cracks.

ϕ_o = Complete elliptic integral which takes account of crack shape (a/2c).

Figure 6 shows a plot of "Y" against crack depth for a crack growing from the toe of a transverse fillet weld from the finite element studies of Hayes and Maddox (1972). Whilst the overall magnification factor, "Y", drops rapidly initially, the non-dimensional stress intensity factor, $K/\sigma\sqrt{\pi B}$, actually increases continuously with increasing crack size.

Fig.6.
(a) Values of $Y = K/\sigma\sqrt{\pi a}$ for cracks at the toes of fillet welds.
(b) Values of K normalised for thickness for cracks at fillet weld toes.

Numerous other analyses have been performed, e.g. for load carrying cruciform fillet welded joints by Frank (1971) and Maddox (1974b), and for non-load carrying fillet welded joints by Burdekin (1981) and Smith (1984).

Stress Analysis of Tubular Nodes

A recent subject for extensive study has been the fatigue behaviour of offshore structures. The stress analysis of fatigue cracks at tubular node intersections, as used in such structures, presents prodigious difficulties. The loadings and geometries are both complex. Numerical analysis requires a full three-dimensional characterisation of the cracked body. As far as the author is aware, this has not yet been attempted for tubular nodes. One interesting aspect of crack growth in these joints is that it tends to be linear for substantial parts of the life, particularly as the crack grows through the thickness, rather than exponential, as one would expect from a simple consideration of the Paris law (Gibstein (1981), and Dover and Dharmavasan (1982)). The reason for this must be that load is shed from the "hot spot" to other areas around the intersection, as the crack at the hot spot grows. An interesting approach, proposed by Dover and Dharmavasan (1982), is to measure the crack growth rate carefully and to use this to predict the instantaneous value of ΔK_{eff} from the crack growth law established on simple specimens. This may seem a round about way of establishing stress intensity factors for tubular nodes, but at present it may be the only way, because of the complexity of a three-dimensional cracked body stress analysis.

OVERALL LIFE PREDICTION

Many authors have used the integration of the crack growth law (Eq.[3]) to predict the fatigue behaviour of welded joints. In some cases such analyses have been used to study the effects of changes in the various parameters, in others the theoretical predictions have been compared with the observed behaviour of welded specimens tested in the laboratory. Only a few can be mentioned here.

General

The present author (1969) showed how widely varying S-N data obtained in a number of separate investigations of butt welded specimens containing lack of root fusion defects could be rationalised using fracture mechanics. This analysis was employed in deriving the simplified approach to defect acceptance levels in BS PD 6493 (see below). Maddox (1974b) was able, very elegantly, to explain S-N data for fillet welded specimens based on the propagation from assumed initial weld toe intrusion depths of 0.1 to 0.4mm. The success of these predictions provided considerable confidence in the approach.

Frank (1971) analysed the fatigue behaviour of cruciform fillet welded joints failing from the weld root and Gurney and Maddox (1973) followed this up by deriving the optimum combinations of weld leg length and penetration to ensure an equal probability of failure from the weld root as from the weld toe.

The effect of thickness

One important outcome from the fracture mechanics analysis of the fatigue behaviour of transverse non-load carrying fillet welds was the prediction that the fatigue strength would tend to decrease with increasing thickness (Gurney, (1979)). This is because "Y", for a given geometry, is a function of a/B. Thus, the high stress region extends further for a thick plate than for a thin one. For two joints with the same initial flaw size, but in different plate thicknesses, the "Y" factor for the thicker plate will be greater. The crack will, therefore, propagate faster, particularly in the small crack stage when most of the life is expended.

The above only applies provided geometrical similarity is maintained. M_k is also a function of the overall width of the attachment from toe to toe. If this is maintained constant, whilst the load carrying plate thickness is increased, the fatigue strength will actually increase with thickness (Burdekin, (1981)).

The two cases are relevant to two different types of practical situation.

In pressure vessels, offshore tubular structures, etc., there will be a tendency for all thicknesses to increase pro rata. In this case, fatigue strength will decrease with increasing thickness. In bridge girders, stiffeners will tend to be of constant thickness, whilst flange thicknesses vary. In this case, fatigue strength will increase with flange thickness.

The fracture mechanics predictions with regard to thickness have been investigated experimentally and reasonable agreement has been established. Cognisance of this effect has been taken in drafting the latest revision of the Department of Energy's Offshore Guidance Notes (Anon, (1984)).

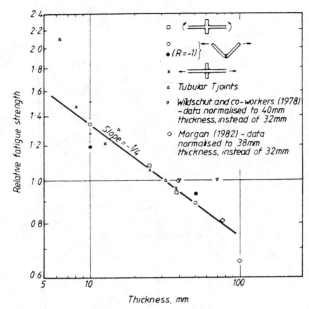

Fig.7. Influence of plate thickness on fatigue strength (normalised to a thickness of 32mm.) (All tests at R = 0 except where stated.) (From Anon, (1984)).

Figure 7 is from the latter and shows the fatigue strength (normalised for a thickness of 32mm) plotted against thickness. From this figure, the following relationship was derived:-

$$S = S_{32}\left(\frac{32}{t}\right)^{\frac{1}{4}} \qquad [11]$$

Thus the fatigue strength of a 100mm thick joint would be about 75% of that of one 32mm thick.

CODES AND STANDARDS

British Standard 5400

The fatigue rules in the bridge code, BS 5400, are based on simple fatigue tests of representative weld details. However, they are strongly supported by fracture mechanics analysis. For example, as mentioned above, Maddox (1974b) showed how the S-N curves obtained from simple tests on specimens with non-load carrying fillet welds could be predicted by integrating the fracture mechanics crack growth law.

This finding has been confirmed more recently by Smith and Smith (1983). The latter authors showed how the fatigue limit for welded joints could be derived from the crack growth law provided that it took account of ΔK_{th}. Thus, BS 5400 acquires implicit support from crack propagation analysis.

British Standard Published Document 6493

British Standard PD 6493 describes a method for assessing the significance of weld imperfections with regard to fitness for purpose. This can be used to replace or supplement existing arbitrary acceptance criteria which are based on good workmanship. It is worth noting that the British Pressure Vessel Code, BS 5500, has a very valuable clause enabling this to be done by agreement between the parties concerned. The use of the more realistic acceptance criteria which can be derived from PD 6493 can save considerable sums by eliminating unnecessary repair work. The present author (1979) has described the derivation of the fatigue clauses in PD 6493 and Wylde and Booth (1981), and Booth and Wylde (1982) have shown how they are applied. Briefly, there are two treatments for planar defects. In one, the "simplified" approach (on first reading the document, this seems the more complicated of the two), the fracture mechanics integration has already been carried out. The user simply has to decide on an effective stress range and life. In other words, he selects a required S-N curve for his structure. This then dictates what size of planar and non-planar defect can be accepted. An advantage of this approach, is that the top six S-N curves are the same as those for Classes C to G in BS 5400. Thus, the designer can, if he chooses, ensure that the quality specified for the welds is as good, in terms of fatigue behaviour as the design detail itself. If there is a fillet weld with a relatively low classification adjacent to a butt weld, there is no point in requiring the latter to have a quality which would secure a much higher fatigue strength.

The second method leaves the user to "do his own thing". He determines an appropriate crack growth law and then integrates this. A step-by-step integration is proposed whereby the stress intensity factor is calculated at the end of each step and this is then assumed to apply for the whole of that step, a conservative assumption. The life is calculated as the sum of the cycles required to propagate the crack through all the steps from the initial flaw size through to failure.

ASME Boiler and Pressure Vessel Code, Section XI

Section XI of the ASME code gives rules for in-service inspection of nuclear power plant components. It contains a non-mandatory Appendix A for the analysis of flaw indications. The procedure is very similar to that set out in PD 6493. The steps listed are as follows:

    1.   Determine ΔK_1, the maximum K_1 range for the initial flaw, for the first transient.

2. Find incremental crack growth, Δa, for this transient from the crack growth law.
3. Update flaw by assuming it grows to a + Δa and remains geometrically similar.
4. Proceed to next transient.

After all the transients have been considered this yields the end of life flaw size, a_f.

The value of K_1 for this flaw size is compared with the fracture toughness and, if it is lower, the original flaw is acceptable.

Figure 8 shows the crack growth relationship given in Appendix A. It will be noted that the "wet" curves for surface breaking defects predict growth rates up to 40 times faster than the air curve for buried defects. The effect of R-ratio is also seen to be significant for the "wet" curves. The growth rate at high R is predicted as being as much as 12 times faster than at low R.

There is one criticism of this procedure, and this is the assumption, made in Step 3, that the flaw remains geometrically similar. In general, the stress intensity factor will vary around the periphery of the flaw. Because of the power law relationship between ΔK and da/dN, the crack will grow much faster at the points around the periphery where the stress intensity is at a maximum. We have seen above how cracks tend to change shape during propagation and how the way in which they change depends on the initial shape and on the type of loading (tension, bending or combined tension/bending). See for example Fig.5.

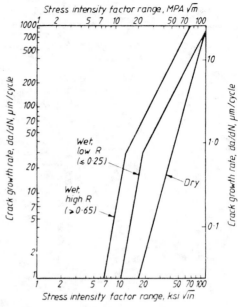

Fig.8. Reference fatigue crack growth curves for carbon and low alloy ferritic steels. (After the ASME Boiler and Pressure Vessel Code, Section XI, (Appendix A)).

CASE HISTORIES

Case histories follow which illustrate aspects of damage tolerance. The first case is a brief description of the failure of the "Alexander L. Kielland".

The next concerns Shell's North Cormorant platform, in which the PD 6493 concepts were applied at a late stage in construction. Finally, the application of PD 6493 as an aid to design is outlined with reference to Conoco's Hutton Tension Leg Platform.

The "Alexander L. Kielland"

The "Alexander L. Kielland" failed by fatigue cracking of a non-redundant brace, 2.6m diameter and 26mm thick from the edge of a 325mm diameter hole, through which a hydrophone support tube had been set in a similar manner to a set-through nozzle in a pressure vessel. The Norwegian Commission of Enquiry (Anon, (1981)) reported that the small fillet welds attaching the support tube to the brace had already partially failed on delivery, paint being found on the fracture surface. They concluded that, at an early stage, the support tube became separated from the brace around three-quarters of the support tube's circumference, so that it ceased to perform its function of reinforcement to the hole. New fatigue cracks then grew from the edge of the hole into the brace finally causing its complete failure.

A very important aspect of this failure is the shape of the crack growth curve. Karlsen and Tenge (1980) carried out a fracture mechanics analysis of the growth of a crack from the edge of the hole. Their results are shown in Fig.9.

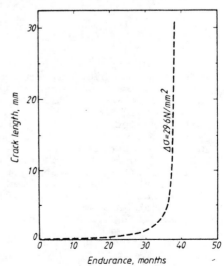

Fig.9. Estimated propagation curve for a crack growing from one side of hydrophone opening. Effective stress range 29.6 N/mm². (After Karlsen and Tenge (1980)).

It will be seen that, according to this analysis, the crack growth rate was relatively slow throughout the first three years of the platform's life. Thereafter, it
accelerated rapidly, final failure taking place after about 3.6 years. Karlsen and
Tenge also plotted their estimates of the crack front position at various stages in
the life. This is shown in Fig.10. It will be seen that their analysis suggests that
the crack would not have developed through the thickness after three years. From
evidence presented in the Norwegian Report one concludes that the crack would only
be reliably detectable by visual means in the last weeks of life. The implications
of this for in-service inspection are obvious. Some degree of structural redundancy
or some form of continuous monitoring must be provided. Clearly the "Alexander L.
Kielland" was neither damage tolerant nor fail safe.

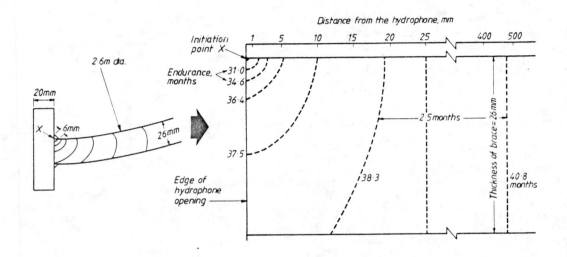

Fig.10. Crack front positions at various stages during the life for an effective
stress range of 29.6N/mm^2. (After Karlsen and Tenge (1980)).

This failure may be contrasted with the partial failure of a fixed offshore platform
(Anon, (1976)). In this case a complete tier of horizontal diagonal bracing parted
company from the legs as a result of fatigue failures at all four brace-to-leg
welds. The failure was only discovered when divers carrying out a routine survey
found this tier of bracing lying across the next tier down! Marine growth on the
fracture surfaces suggested that it had been there for at least six months. This
structure was clearly a fail-safe design.

North Cormorant Jacket

At a late stage in fabricating the North Cormorant offshore platform, chevron cracks
(weld metal hydrogen cracks) were found in submerged-arc longitudinal and girth
seams in the tubulars used to make the large intersections (nodes). The tubulars,
which were up to 100mm thick, had been made by one subcontractor. A second subcontractor had welded brace stubs to these to make the nodes which were then postweld
heat treated. The main contractor had received the nodes and welded them into the
jacket. The chevron cracking was first found during ultrasonic testing (UT) of one
of the circumferential erection welds, when the NDT Technician had to probe through
the end of the longitudinal seam of the node, where this intersected the erection
weld. Further UT probing revealed that 20% of seam welds were affected. Repair of
the defects would have had horrendous consequences for the construction programme.
The structure would have had to be dismantled, because in-situ PWHT of the repair
welds was impossible. It was therefore decided to carry out an assessment of the
cracks using PD 6493 procedures. Fortunately the welding procedures were known to
have excellent toughness. The defects were assessed for resistance to fracture and
fatigue and found to be innocuous. It was decided, with the concurrence of the Cer-
tifying Authority, to allow construction to proceed.

The value of this decision can only be estimated, but it can be assumed that, without it, delivery would have been delayed for at least 12 months, because launch in
the summer weather window would have been missed. With a total investment in the

North Cormorant field of £325M, the loss in interest charges alone, at rates then prevailing, would have been about £45M.

The Hutton Tension Leg Platform

Smith and co-authors (1982) describe the damage tolerant design of the deck part of the vessel structure for Conoco's TLP. The overall design approach was based on the limit states concept, one of the limit states being fracture. Fracture mechanics was used to assess the condition for final instability and a crack growth analysis was carried out to ensure that any hypothetical weld discontinuity would not grow to this condition within the design life of the platform. The analysis took account of the crack growth threshold. Typical weld toe cracks were modelled using finite elements. As an example, a 1.0mm deep crack at the toe of a highly stressed butt weld in 40mm thick plate was predicted to grow to a depth of nearly 6mm in the twenty year life. This was acceptable provided that the fracture toughness exceeded a crack tip opening displacement of 0.15mm. The fracture mechanics analysis provided a valuable back up to the main design approach which used the S-N curves published in the Department of Energy's Guidance Notes.

CONCLUDING REMARKS

This paper has set out to show the contribution which crack growth analysis, in part founded on the work of the three scientists honoured by this volume, now makes to the assessment of the significance of discontinuities. The background has been welding, the author's area of interest; but the principles are more general and assessment methods similar to those described here are being applied in a wide range of industrial situations, often with considerable financial benefit.

Each of the main factors in a fracture mechanics crack growth analysis are discussed. The importance of well written and updated codes cannot be over emphasied. The case histories are used to illustrate the application of crack growth analysis in three typical situations. A failure analysis, the assessment of known imperfections found during construction and as an adjunct to normal design to assess the size of hypothetical defect which might lead to failure.

ACKNOWLEDGEMENTS

The author is grateful to his colleagues at The Welding Institute, particularly Dr. S.J. Maddox and Mr. P.J. Mudge, for helpful discussions whilst writing this paper.

REFERENCES

Anon. (1976) "Partial failure of a fixed offshore platform". Metal Construction, 8, 312-314.
Anon. (1979) "Reports of the Plate Inspection Steering Committee (PISC) Exercise", Report No. EUR 6371, Vols I-V, Commission of the European Communities Nuclear Science and Technology Services, Luxembourg.
Anon. (1981) "The 'Alexander L. Kielland' accident", Norwegian Public Reports, NOU 1981: 11.
Anon. (1984) "Background to new fatigue guidance for steel welded joints in offshore structures", HMSO, London.
ASME Boiler & Pressure Code, (1983), Section XI, "Rules for inservice inspection of nuclear power plant components".
Austen, I.M. (1979) "An analysis of the applicability of empirical fatigue crack growth relationships", British Steel Corporation, Res. Rep. SH/PT/6795/9/79/B.
Bartle, P.M. and Mudge, P.J. (1983) "Acoustic pulsing - a technique for remote defect monitoring", Welding Inst. Res. Bull., 24, 292-294.
Beevers, J.C. and Carlson, R.L. (1985) "Considerations of the significance of factors controlling fatigue thresholds". [This volume].
Bennett, A. and Brown, D. (1983) "Inspection of underwater structures by flooded member detection", Brit. J. of Non-Destructive Testing, 25, 121-123.
Booth, G.S. and Wylde, J.G. (1982) "Defect assessment for fatigue - example solutions using the techniques described in PD 6493", Welding Inst. Res. Bull, 23, 84-88.
BS 5400:Pt 10:1980. "Steel, concrete, and composite bridges; code of practice for fatigue", The British Standards Institution, 1980.
BS 5500:1985. "Unfired fusion welded pressure vessels", The British Standards Institution, 1985.
BS PD 6493:1980 "Guidance on some methods for the derivation of acceptance levels for defects in fusion welded joints", British Standards Institution.
Burdekin, F.M. (1981) "Practical aspects of fracture mechanics in engineering design", Proc. I Mech E, 195, 73-86.

Clayton, A.M. and Morgan, H. (1984) "The development of cracks in welded connections", IIW Document XIII-1132-84.

Cullen, W.H. and Torronen, K. (1980) "A review of fatigue crack growth of pressure vessel and piping steels in high temperature, pressurized reactor grade water", Naval Res. Labs., Washington D.C., Report No. NRL-MR-4298.

Davenport, R.T. (1982) "A review of fatigue crack growth laws and some recommendations for calculating residual fatigue life", CEGB Report No. NW/SSD/SR/69/81.

Dover, W.D. and Dharmavasan, S. (1982) "Fatigue fracture mechanics analysis of T and Y joints", Offshore Technology Conf., Houston, OTC 4404.

Frank, K.H. (1971) "The fatigue strength of fillet welded connections", PhD Thesis, Lehigh University.

Frost, N.E. (1959) "A relation between the critical alternating propagation stress and crack length for mild steel", Proc. I Mech E, 173, 811-836.

Frost, N.E., Pook, L.P. and Denton, K. (1971) "A fracture mechanics analysis of fatigue crack growth data for various materials", Engineering Fracture Mechanics, 3, 109-126.

Garwood, S.J. (1979) "Fatigue crack growth threshold determination", Welding Inst. Res. Bull., 20, 262-265.

Gibstein, M.B. (1981) "Fatigue strength of welded tubular joints tested at Det norske Veritas Laboratories", Int. Conf. on 'Steel in Marine Environments', Commission of the European Communities.

Gurney, T.R. and Maddox, S.J. (1973) "Proposed fatigue design stresses for weld metal", Welding Inst. Report E56/73.

Gurney, T.R. (1979) "The influence of thickness on the fatigue strength of welded joints", 2nd Int. Conf. on Behaviour of Offshore Structures (BOSS 79), London.

Harrison, J.D. (1969) "The analysis of fatigue test results for butt welds with lack of penetration defects using a fracture mechanics approach", Fracture 69, Proc. 2nd Int. Conf. on Fracture, Chapman & Hall.

Harrison, J.D. (1970) "An analysis of data on non-propagating cracks on a fracture mechanics basis", Metal Construction, 2, 93-98.

Harrison, J.D. (1979) "Acceptance levels for defects in welds subjected to fatigue loading", IIW Colloquium on Practical Application of Fracture Mechanics, Bratislava.

Hayes, D.J. and Maddox, S.J. (1972) "The stress intensity factor of a crack at the toe of a fillet weld", Welding Inst. Res. Bull., 13, 15-17.

Iida, K. (1980) "Shapes and coalesence of surface fatigue cracks", Proc. Int. Symp. on Fracture Mechanics, Beijing, China.

Iida, K. (1983) "Aspect ratio expressions for part-through fatigue cracks", IIW Document XIII-967-80.

James, L.A. and Knecht, R.L. (1975) "Fatigue crack propagation behaviour of type 304 stainless steel in a liquid sodium environment", Metal Trans. A., 6, 109-116.

Jaske, C.E., Payer, J.H. and Balint, V.S. (1981) "Corrosion fatigue of metals in marine environments", US Dept. of Commerce, Washington D.C.

Jessop, T.J., Mudge, P.J., Harrison, J.D., Charlesworth, J.O., Aldridge, E.E., Silk, M.G., Coffey, J.M., Bowker, K.J., Wrigley, J.M. and Denby, D. (1982) "Size measurement and characterisation of weld defects by ultrasonic testing; Part 2: Planar Defects", Welding Inst. Report Series 3527/10/80.

Karlsen, A. and Tenge, P. (1980) "D/R 'Alexander L. Kielland', Estimation of fatigue cracks in hydrophone region on brace D-6", Det Norske Veritas Report 80-0410.

Knott, J.F. (1985) "Models of fatigue crack growth". [This volume].

Lindley, T.C. and Nix, J.K. (1984) "Metallurgical aspects of fatigue crack growth". [This volume].

Maddox, S.J. and Webber, D. (1978) "Fatigue crack propagation in aluminium-zinc-magnesium alloy fillet welded joints", Fatigue Testing of Weldments, ASTM-STP-468, D.W. Hoeppner (Ed).

Maddox, S.J. (1974a) "Fatigue crack propagation data obtained from parent plate, weld metal and HAZ in structural steels", Welding Res. Int., 4, 36-60.

Maddox, S.J. (1974b) "Assessing the significance of flaws in welds subject to fatigue", Welding J., 53, Res. Suppl., 401s-409s.

Maddox, S.J. (1975) "An analysis of fatigue cracks at fillet welded joints", Int. J. of Fracture, 11, 221-243.

Morgan, H.G. (1982) "The effects of plate thickness on the fatigue performance of simple welded joints", UKAEA Northern Division Report ND-R-941(S), HMSO.

Mudge, P.J. and Williams, S. (1981). "Performance of ultrasonic non-destructive methods for defect assessment in relation to fracture mechanics requirements", Conf. on Fitness for Purpose Validation of Welded Constructions. The Welding Institute, London.

Paris, P.C. and Erdogan, F. (1963) "Critical analysis of crack propagation laws", Trans. J. Basic Eng., 85, 528-34.

Pearson, S. (1966) "Fatigue crack propagation in metals", Nature, 211, 1077.

Pook, L.P. and Frost, N.E. (1973) "A fatigue crack growth theory", Int. J. of Fracture Mechanics, 9, 53-61.

Richards, C.E. and Lindley, T.C. (1972) "The influence of stress intensity and microstructure on fatigue crack propagation in ferritic materials", Engineering Fracture Mechanics, 4, 951-978.

Signes, E.G., Baker, R.G., Harrison, J.D. and Burdekin, F.M. (1967) "Factors affecting the fatigue strength of welded high strength steels", Brit. Weld. J., 14, 108-116.

Silk, M.G. (1979) "Defect sizing by ultrasonic diffraction", Brit. J. of Non-Destructive Testing, 21, 12-15.

Smith, I.F.C. and Smith, R.A. (1982) "Defects and crack shape development in fillet welded joints". Fatigue of Engineering Materials and Structures, 5, 151-165.

Smith, I.F.C. and Smith, R.A. (1983) "Fatigue crack growth in a fillet welded joint", Engineering Fracture Mechanics, 18, 861-869.

Smith, I.J. (1984) "The effect of geometry changes upon the predicted fatigue strength of welded joints", Proc. 3rd Conf. on Numerical Methods in Fracture Mechanics, Swansea.

Smith, I.J., Pisarski, H.G., Ellis, N. and Prescott, N.J. (1982) "Assessment of fracture toughness requirements for the deck structure of the Hutton TLP using finite elements and the CTOD design curve", 14th Offshore Technology Conference, Houston, OTC. 4430.

Sommer, E., Hodulak, L. and Kordisch, H. (1977) "Growth characteristics of part-through cracks in thick walled plates and tubes", J. Pressure Vessel Technology, 2, 106-111.

Speidel, M.O. (1982) "Advances in fracture research", Vol. 6, (D. François, Ed.), Pergamon Press, Oxford. 2685-2704.

Walker, E.F. (1981) "Effects of marine environment", Int. Conf. on Steel in Marine Structures, Commission of the European Communities.

Watkinson, F, Bodger, P.H. and Harrison, J.D. (1970) "The fatigue strength of welded joints in high strength steels and methods for its improvements", Proc. Conf. Fatigue of welded structures, The Welding Institute.

Wildschut, H., de Back, J., Dortland, W. and van Leeuwen, J.L. (1978) "Fatigue behaviour of welded joints in air and sea water", European Offshore Steels Research Seminar, The Welding Institute.

Williams, S. and Mudge, P.J. (1982) "Fatigue crack growth monitoring: Fracture mechanics and non-destructive testing requirements", Conf. on Periodic Inspection of Pressurised Components, Instn. of Mech Eng, London.

Wylde, J.G. and Booth, G.S. (1981) "Defect assessment for fatigue - the technique described in PD 6493", Welding Inst. Res. Bul., 22, 91-97.

Crack Growth — Present Status, Future Direction

B. TOMKINS

UKAEA

ABSTRACT

This paper identifies types of problems to which the advances in our understanding of fatigue crack growth can be applied. Several key areas of research are suggested; progress in which should enable a wider use of our fracture knowledge to be made.

KEYWORDS

Fatigue; fatigue cracks, fatigue crack growth; crack growth threshold; non-destructive testing.

INTRODUCTION

Fatigue in metals is a failure process which we now understand. It involves the development of a dominant crack, whose size increases with cycles of applied stress or strain until the material section is so weakened that it breaks. The fact that the repeated applied stress could be well below that needed to cause yielding, let alone failure, combined with limited observation of cracks prior to failure were the main reasons for our slowness in understanding the process. The success of Peter Forsyth and Gerry Smith in revealing the mysteries of fatigue was due to their persistence in observing the process on the right scale. Their painstaking micro-scopic studies showed how crack initiation could occur at a free surface as a result of slip band development, even below general yield, and subsequent crack growth occur by incremental localised failure at the crack tip. As with any physical pro-cess, knowledge of the mechanism is a prerequisite for confidence in its quantifica-tion. Norman Frost's experiments investigated crack development on an engineering rather than microscopic scale, but he sought to quantify the rate of crack growth rather than its integration, namely life, which until then had been the only concern of engineers. When combined, these and subsequent studies of crack scale and crack growth rate provide the basis for our confidence in understanding the fatigue process as it operates in engineering components and structures.

In 1980, the Royal Society sponsored a discussion meeting on Fracture Mechanics in Design and Service, subtitled 'Living with Defects'. This timely meeting reminded us that during the last decade we have come to accept that flaws may exist in our structures both as pre-existing defects and cracks which have developed in service usually as a result of some form of fatigue. We are now faced with putting our understanding of the fatigue process to practical use by providing engineers with rules by which the rate of growth of cracks in service can be estimated. This 'damage tolerance' approach to the assurance of continuing structural integrity is used in an increasing number of areas of application ranging from aero-engines and nuclear power plant to large offshore structures. However, the fracture mechanics tools needed for the assessment of crack growth in service are somewhat limited at

133

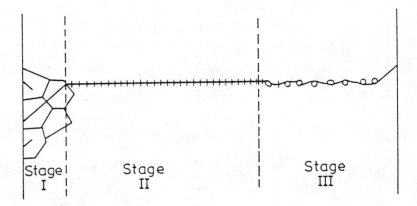

Fig. 1: Stages of fatigue crack propagation across a specimen section

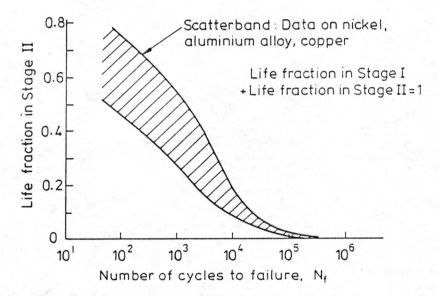

Fig. 2: Fractions of life spent in Stage II crack growth (after Laird and Feltner, 1967)

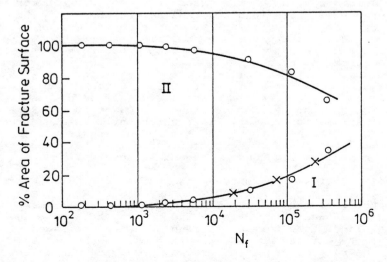

Fig. 3: The relative areas of fracture surfaces occupied by Stages I and II growth (after Laird and Smith, 1962)

present. The future for crack growth studies in fatigue lies in the development of these tools for a widening range of applications. This will involve studies across the whole range of scale of cracks from microns to metres and growth rates from atomic dimensions to millimetres per cycle.

THE SCALE OF FATIGUE CRACKS

Lack of appreciation of the scale of fatigue cracks has been the main source of misunderstanding by the engineer of the fatigue process. Naturally initiated fatigue cracks on a smooth polished surface are of order 1 - 10 μm. This is the size at which in many metals and alloys they behave as cracks under control of the continuum response of the material rather than its local microstructure. In the more sophisticated high strength alloys, local microstructure can control crack behaviour well beyond this point, but in most materials by the time a crack is of the order 100 μm (0.1 mm) its growth is controlled solely by continuum response. For many materials, crack growth follows three stages (Fig. 1), the first two of which were noted by Forsyth in his classic paper of 1962. His observations on aluminium alloys where on such a fine scale that discrete development of cracks of micron dimensions was clearly seen. The first stage (I) has been shown to be a continuation of the crack initiation process following dislocation sub-structure development. Even in this stage, continuum behaviour can dominate growth, particularly if the substructure is well developed throughout the grain in which the crack starts. However, stage I growth is the stage most susceptible to the influence of local microstructure and local dislocation movement. From observations on pure nickel and aluminium, both high stacking fault energy FCC metals which form a tight dislocation cell structure, Laird and Feltner (1967) showed that the proportion of life spent in stage I increased sharply and the proportion in stage II decreased as fatigue life itself increased for lower stress cycling, see Fig. 2. Rather earlier, Laird and Smith (1962) noted that this increase in stage I life was not accompanied by a comparable increase in fracture surface area created by stage I growth. So in high cycle, low stress fatigue, where stage I growth dominates, crack growth rates are very slow and crack sizes therefore remain small. This accounts for the inability to see cracks on an engineering scale until late in life, say 80 - 90%. The detection of naturally initiated fatigue cracks in engineering components prior to failure remains a major problem today, despite advances in non-destructive examination techniques. It requires continuous monitoring of components at the potential crack location, which, at the sensitivity required, is usually either impossible or very expensive and unreliable. If naturally initiated fatigue cracks are detected in service, it usually indicates that their rate of growth has slowed down, a point which will be enlarged upon later.

The second stage of crack growth (II) also identified by Forsyth is characterised by a growth direction normal to the applied maximum principal stress or strain range, see Fig. 1. It is often associated on the fracture surface by discrete striation markings each indicating a successive position of the crack front. Under a constant applied cyclic stress or strain field, the width of a striation in the direction of crack advance is proportional to the current crack size. Striations have been observed over a wide range of sizes, being as small as 0.1 μm, in a very ductile alloy, up to several microns. As will be seen later, the larger the striation size, the faster the crack growth rate. However, striation size does not necessarily correspond to crack growth increment per cycle, it often taking several cycles to advance the crack by one striation. In fact, striations are produced by the intense plastic flow at the crack tip involved in crack advance (Laird and Smith, 1962; Tomkins, 1968). Stage II crack growth is considerably faster than stage I and accounts for most of the fracture surface produced by a fatigue failure.

At high strain levels, Laird and Smith (1962) observed that cracks develop immediately as stage II Cracks, thus bypassing the stage I process. The short lives observed in high strain fatigue are a result of the dominance of the faster stage II growth in these failures. Stage II growth can also dominate the development of long cracks at low applied cyclic stress levels but again the rate of growth will be relatively fast, being a function of crack size. Observations of both short and long crack growth over the past 30 years have established that although the scale of crack can vary enormously, from microns to metres, the mechanism of crack advance is the same, be it by the slow stage I process or the faster stage II process. This similarity underpins our confidence in predicting the rate of fatigue crack growth in terms of applied stress/strain range and crack size.

THE RATE OF GROWTH OF FATIGUE CRACKS

Fast and slow are relative terms for crack growth rate and in fact the rate can be related directly to the applied stress or strain range. Frost (1959) was one of the first to examine systematically the dependence of crack growth rate (da/dN) on applied stress amplitude $\Delta\sigma$, in tests on thin cracked plates containing a relatively long (several mm) centre crack of length 2a. He discovered that da/dN was proportional

138

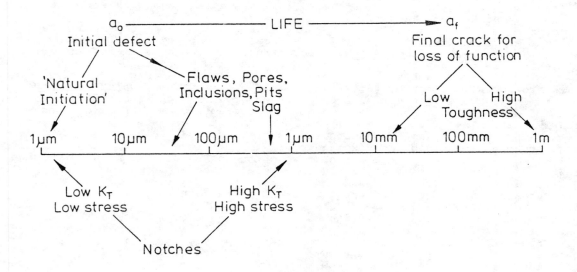

Fig. 5: Various initial and final crack sizes defining
useful crack propagation life.

where f(a) is dependent on the stress amplitude ($\Delta\sigma$) through a relationship with a described by ΔK or ΔJ, e.g. equation (1). For a given material, fatigue prone combinations of stress amplitude and crack sizes can be defined if the limiting K parameters for crack growth, ΔK_{TH} and K_C are known. The bounding curves show crack sensitive and crack insensitive regions. An increase in material strength accompanied by a reduction in material toughness lowers the threshold crack size for sensitivity even though it increases the basic material fatigue strength below this limit.

In the following sections, these parameters are considered in detail, noting the status of present knowledge and the likely direction of future developments. It is clear that although the importance of these four parameters to any assessment has been established, specific knowledge of each is required for specific applications. In this sense the fracture mechanics approach is similar to the S-N approach to fatigue failure, where S-N curves must be chosen carefully for each application.

INITIAL CRACK SIZE (a_o)

Figure 5 indicates the wide range of initial crack sizes encountered in structures and components. Naturally initiated fatigue cracks in clean metals and alloys are of the order 1 - 10 μm, a size similar to that of inclusions and second phase particles in many practical alloys. Other material flaws encountered in fabrications are over a wider range and where surface breaking, can provide ready made crack starters. These include porosity in castings, pores, microcracks, lack of fusion and slag intrusions in welds, which can range up to millimetre dimensions in large structures or components. Environmental attack resulting in pitting, or inadvertent mechanical surface damage can also be considered as flaws of up to millimetre dimensions.

Now, as the naturally initiated fatigue crack size is of the order 1 - 10 μm, microstructural or structural flaws often exceed this and therefore have the potential to significantly reduce the fatigue life of a component if they are in a fatigue prone location and behave from the outset as cracks. Equation (2) can be used to estimate the effect of initial crack size (a_o) as a function of final crack size (a_f). Using a crack growth law of the form of equation (1) with ΔK exponents of 2, 3 and 4, Table 1 shows the effect on life of a two order of magnitude increase in initial crack size, say from 10 μm to 1 mm for two final crack size conditions of 10 and 100 mm.

Table 1:

Life	Initial crack a_o	Final crack a_f	a_f/a_o	Life ratio, N_1/N_2 m = 2	m = 3	m = 4
N_1	10 µm	10 mm	10^3			
				3	14	111
N_2	1 mm	10 mm	10			
N_1	10 µm	100 mm	10^4			
				2	11	101
N_2	1 mm	100 mm	10^2			

Although these simple calculations do not include the effects of crack size on compliance in the definition of ΔK, they show that for typical values of 3 or 4, the presence of a 1 mm initial crack readily reduces the material fatigue life by one or two orders of magnitude. For short cracks, at higher strain levels, m tends to be ~ 2, i.e. crack growth is exponential, and the effect of initial crack size is less significant.

The potential sensitivity of life to a_o makes it a critical parameter for any fracture mechanics based life assessment of a component or structure. The effective a_o must be known for the fatigue prone location and, for components whose failure would be catastrophic, an inspection interval must be determined which enables an ongoing assurance of integrity to be provided. Actual knowledge of a_o depends on an initial non-destructive examination (NDE). Current methods for surface examination utilising ultrasonic, eddy current or electric potential drop techniques can be sensitive enough to indicate flaws of order several µm under favourable circumstances. However, for large constructions this sensitivity may be reduced to several mm and in adverse circumstances to tens of mm. Often, the NDE sensitivity itself is used to define a maximum flaw size which could be present, all larger detected flaws being repaired or removed. In this case, the detectable flaw so defined (a_d) is equal to or greater than the actual initial flaw size a_o. It is the size which must be used to fix the inspection interval in terms of cycles, or time, based on knowledge of the likely crack growth rate, using a relationship such as equation (1) and acceptable final crack size, a_f. Equation (2) can again be used to show the interrelation between a_o, a_d and the inspection period. The number of cycles, N for a crack to grow from a to xa is given by,

$$N = \frac{a}{(da/dN)_a} \cdot \frac{1}{(m/2 - 1)} \cdot \left[1 - \frac{1}{x^{(m/2-1)}} \right] \quad \text{for } m \neq 2 \quad (3)$$

If x represents the ratio of $a_d:a_o$ then N is the smallest cyclic interval for which crack detection can be assured. a_f/a_d must also be at least of order x if the use of N as an inspection interval has a chance of detecting a crack prior to failure. Periodic in-service inspection monitoring of fatigue crack growth can be successful only if a_o is close to a_d. Otherwise, cracks are either unlikely to be detected at all or will produce failure between inspections for anything other than low cyclic frequency fatigue situations.

Because of the sensitivity of fatigue life to initial crack sizes of 1 mm or less, and the difficulties of reliably detecting such cracks non-destructively, it is sensible to safeguard life by the prevention or removal of such cracks. Detailed, accurate stress analysis is readily repaid if it identifies fatigue prone regions and enables careful surface detailing and finishing to be applied in these areas. Consideration of service history must also identify potential causes of initial cracks by means other than fatigue itself. Corrosion pitting, fretting and scoring during maintenance have all led to premature failure because their importance as crack initiators was overlooked. In this regard, the studies of cracks and their growth have provided new evidence for the well established, but still often ignored, surface factors which can precipitate fatigue failure in real components.

FLAWS OR CRACKS?

Surface flaws on the sub-millimetre scale occur so readily that it is important to ask whether such flaws inevitably behave as cracks. All flaws, (such as scores, pits, pores, etc.) including cracks, are effective in fatigue because they are stress concentrations. The difference between a crack and a simple stress concentrator is that under load, the former will extend fractionally because the associated stress concentration produces plastic deformation so localised that a microscopic element of new crack surface is produced. Less severe flaws will induce more general plastic deformation. This can provide a site for crack initiation by local higher strain fatigue, but the effect on overall fatigue life is more limited.

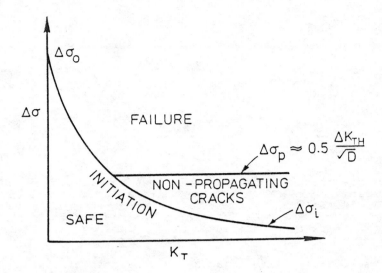

Fig. 6: Failure regimes in notched components (after Frost and Dugdale, 1957)

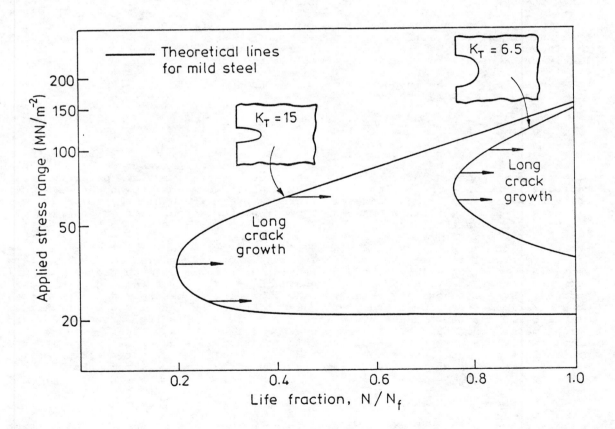

Fig. 7: Life fraction at which long crack growth begins as a function of stress concentration factor (after Cameron and Smith, 1981)

Figure 6 reproduces a plot, of the type suggested by Frost and Dugdale (1957), of fatigue strength vs notch elastic stress concentration factor (K_T) for macroscopic notches. As K_T increases, the fatigue strength is reduced as $1/K_T$ by a simple effect of the stress concentration in providing a location for the material fatigue strength $\Delta\sigma_0$ to be exceeded and a crack initiated. For sharp notches, a higher stress is needed to propagate the initiated crack to failure. Cameron and Smith (1981) have studied the fraction of life spent in the initiation and growth of cracks in a low strength structural steel from notches with various K_T's under various levels of applied cyclic stress. Figure 7 summarises their results for a high and low K_T notch. As K_T increases, the fraction of life spent in long crack growth increases, crack initiation and growth within the notch root region being minimal over a considerable band of applied stress range. It was also found that life itself at a given applied stress range decreased considerably with increase in the amount of long crack growth. Together, Figs. 6 and 7 show the strong influence of K_T in overcoming the crack initiation barrier and readily creating a large effective a_0 under fatigue loading. Similarly, the application of high stresses to lower K_T notches can readily initiate cracks by high strain fatigue. The two effects can be summarised schematically in the plot of Fig. 5. High K_T factors and high stresses promote the early formation of relatively large initial cracks whilst low K_T factors and low stresses encourage smaller effective cracks to form.

Now most incipient flaws are micro- rather than macroscopic stress concentrations, but their response to an applied stress will be similar, provided that uniform plastic deformation can occur. For behaviour as a crack, the root radius (ρ) would have to be very small, of the order of typical fatigue crack opening displacements (δ), say < 1 μm, whereas it is often likely to be a small fraction of the flaw size (D). Now the stress concentration factor K_T is $\cong 2\sqrt{D/\rho}$, so many flaws will not have sufficient D/ρ aspect ratio to rapidly produce cracks of effective size D. The most effective crack forming microflaws are those which have at some point on their circumference a small root radius ρ. These can be most readily formed by microstructural damage processes such as fretting or corrosion rather than surface mechanical damage. Figure 8 shows how corrosion pitting in an austenitic steel gives fatigue endurance behaviour consistent with an integrated long fatigue crack growth law, with a_0 given by the order of the pit depth.

Again, consideration of typical surface flaws as crack starters encourages study of the type and scale of flaw before taking the simple assumption that every flaw is a crack from the outset. It can often take most of the fatigue life to convert a flaw

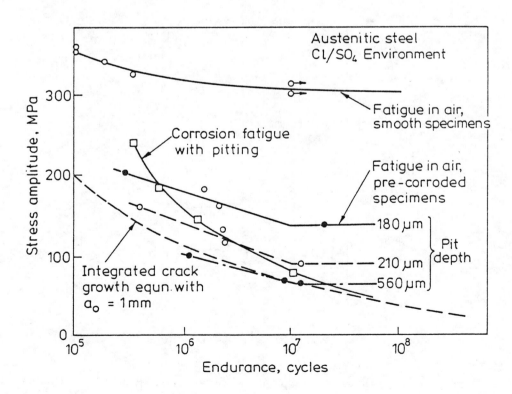

Fig. 8: Effect of pitting on the fatigue strength of an austenitic steel

144

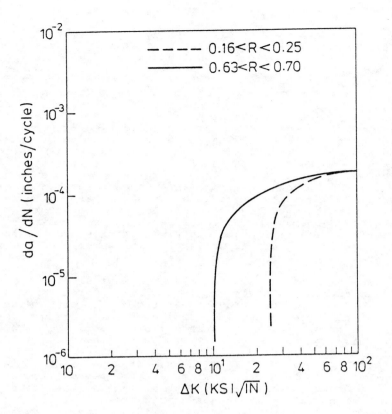

Fig. 9: Effect of stress ratio ($R = \sigma_{min}/\sigma_{max}$) on fatigue
crack growth rate in PWR environment (1 cycle per min)
(adapted from Bamford, 1977)

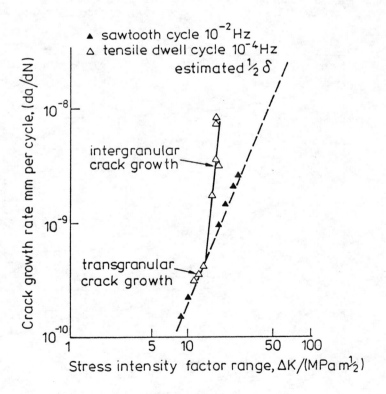

Fig. 10: Effect of creep on fatigue crack growth in 316 stainless
steel at 625°C, R = 0.6 (after Lloyd and Wareing, 1979)

conditions, this has the effect of increasing the aspect ratio by promoting surface growth at the expense of through thickness growth. Fatigue always involves the development of a dominant crack and, where several cracks are initiated or present initially, crack linkage will occur in such a way as to develop a major crack consistent with the stress field which pertains. It might also be noted that stress intensity factor solutions for this type of crack are difficult to obtain with recourse to cumbersome 3-D finite element calculations.

Crack shape development studies and its prediction are strong research topics at the present time as shape development ultimately determines both the time to failure and form of failure in a structure. For example, in relation to time to failure, a thumbnail crack in bending develops at an approximately linear rate through the section whereas a surface line crack would go through almost exponentially. However, with regard to final failure, a high aspect ratio crack developed by bending is more likely to induce fast fracture on section breakthrough than a low aspect ratio tensile crack.

In their papers in this volume, Lindley and Harrison have noted a number of metallurgical and mechanical factors influencing fatigue crack growth. These effects are usually not very strong in the mid-ΔK region away from the threshold and crack instability points. However, the simultaneous operation of a second damage process in addition to fatigue itself can lead to more dramatic effects which are difficult to predict. These include effects of an aggressive environment and creep damage at elevated temperature.

Environmental or corrosion fatigue concerns the growth of fatigue cracks in an environment, usually aqueous, which is active with respect to the fresh metal surface created at the crack tip to such an extent that it produces additional crack increment. This requires time for a significant effect and it is most often found during low frequency cycling. In the limit, crack extension becomes an incremental stress corrosion process by either the slip-dissolution process or hydrogen cracking on a microscale.

Figure 9 shows data, Bamford (1977), for a pressure vessel steel tested in a controlled reactor water environment at low frequency (0.01 Hz). Although this material is not susceptible to sustained stress corrosion cracking in such an environment, the synergetic effect of fatigue and crack tip dissolution produce accelerated crack growth over a significant region, particularly at a high R ratio. Because corrosion fatigue effects involve a matching of the mechanics and chemical kinetics, they are difficult to predict and require considerable testing effort to bound, because of the need to test at low frequencies. This is a departure from the traditional reliance on high frequency testing in fatigue to provide design and assessment data. High strength alloys are particularly susceptible to environmental effects on crack growth, even if environments are benign as regards simple surface attack.

Another synergism can occur during fatigue crack growth at elevated temperature if this is sufficient to produce creep cavitation damage in the material. Low frequency cycling and intermittent dwell periods under load can induce cavitation within the plastic zones ahead of a crack, particularly in low creep ductility materials. This can subsequently interact with the growing fatigue crack to produce additional crack advance by incremental tearing. Figure 10 shows data, Lloyd and Wareing (1979), for Type 316 stainless steel tested at 625°C under a cyclic frequency of 10^{-4} Hz including a tensile dwell period. Crack growth by incremental tearing through cavitated grain boundaries is observed at a very low level of ΔK. This is a specific example of creep-fatigue interaction which illustrates that a strong effect on growth rate can be obtained. Again only testing at very low frequency revealed this effect.

As crack growth data is applied to more and more practical situations, the possibility of time dependent effects whether by corrosion or creep must be considered. Crack growth testing can be done in 'real time' as far as a service cyclic history is concerned and the amount of such testing will increase in the future. With advanced testing techniques, it is possible to build up a composite picture of crack development in a component or structural detail by elements of such real time testing on realistically flawed specimens. This provides the best means of data extrapolation to plant life times, particularly for long life plant operating in difficult environments.

FINAL CRACK SIZE (a_f)

Some indication of the final tolerable size (a_f) of a growing crack in a structure is essential for the use of a fracture mechanics assessment. For low toughness materials, this is easy to estimate from small specimen toughness data. However, many structural alloys have significant fracture resistance and a tolerable crack size is less easy to estimate. Note must be taken of the crack location and shape as 'tolerable' may refer to distortion or leakage rather than fast, unstable fracture.

Crack growth data on the approach to instability is also difficult to obtain for higher toughness material as it requires the testing of large specimens. There is some evidence, Tomkins (1980), that the fatigue crack growth curve merges with the static fracture resistance curve at large crack increments. If this is confirmed, the generation of relevant high rate data will be simplified.

SUMMARY

The past thirty years have produced a wealth of data on and understanding of fatigue crack growth. However, the accurate application of this type of data and understanding to engineering fatigue problems is not easy and in many cases should not be attempted. It is good that we now accept that the structures we build and components we manufacture are flawed, but where fatigue is involved, this should not automatically lead to a damage tolerance assessment using fracture mechanics. A perfect assessment is no substitute for better quality control of material, processing and fabrication where this is possible and economic. The S-N approach is often more appropriate, particularly for redundant structures and high cycle circumstances.

However, the fracture mechanics assessment route is available and does provide greater assurance of integrity if used correctly. In future it will be used more widely for the reasons noted earlier. We must therefore continue to advance the method by focussed research in a number of areas. These include:

- Crack detection, both pre- and in-service

- Stress and crack analysis, using computer methods to provide tools which can be readily used in design procedures

- Flaw characterisation, in terms of stress concentrations or cracks

- Threshold studies

- Crack growth studies, particularly in realistic environments

The challenge for the future lies in the practical use we make of our knowledge of fatigue crack growth.

REFERENCES

Bamford, W H (1977). The effect of pressurized water reactor environment on fatigue crack propagation of pressure vessel steels. In The Influence of Environment on Fatigue, I.Mech.E., London, 51-56

Cameron, A D and R A Smith (1981). Upper and lower bounds for the lengths of non-propagating cracks, Int. J. Fatigue, 3, 9-15

Dowling, N E (1976). Geometry effects and the J-integral approach to elastic-plastic fatigue crack growth. In J L Swedlow and M L Williams, Cracks and Fracture, ASTM STP 601, 19-32

Forsyth, P J E (1962). A two stage process of fatigue crack growth. In Proc. Crack Propagation Symposium, Vol. 1, Cranfield College of Aeronautics, 76-94

Frost, N E (1959). Propagation of fatigue cracks in various sheet materials, J. Mech. Eng. Sci., 1, 151-170

Frost, N E and D S Dugdale (1957). Fatigue tests on notched mild steel plates with measurements of fatigue cracks, J. Mech. Phys. Solids, 5, 182-192

Frost, N E, L P Pook and K Denton (1971). A fracture mechanics analysis of fatigue crack growth data for various materials. Eng. Fract. Mech., 3, 109-126

Laird, C and C E Feltner (1967). Coffin-Manson law in relation to slip character. Trans. Met. Soc., AIME, 239, 1074-1083

Laird, C and G C Smith (1962). Crack propagation in high stress fatigue, Phil. Mag. 7, 847-857

Lindley, T C (1984). Near threshold fatigue crack growth: experimental methods, mechanisms and applications. In L H Larsson, Subcritical crack growth due to fatigue stress corrosion and creep, Elsevier Applied Science Pub., London, 167-213

Lloyd, G J and J wareing (1979). Stable and unstable fatigue crack propagation during high temperature creep-fatigue in austenitic steels: the role of precipitation. Trans. ASME, J. Engng. Mater. Tech., 101, 275-283

Paris, P C and F Erdogan (1963). Critical analysis of crack propagation laws. Trans. ASME, J. Basic Eng., 85, 528-534

Tomkins, B (1968). Fatigue crack propagation - an analysis, Phil. Mag., 18, 1041-1066

Tomkins, B (1975). The development of fatigue crack propagation models for engineering applications at elevated temperatures, Trans ASME, J. Eng. Mats and Tech., 97, 289-297

Tomkins, B (1980). Micromechanisms of fatigue crack growth at high stress. Met. Sci., 14, 408-417

Tomkins, B (1981). Subcritical crack growth: fatigue, creep and stress corrosion cracking. Proc. Roy. Soc., Series A, 299, 31-44